Cambridge International
AS & A Level Mathematics:
Mechanics

Practice Book

CAMBRIDGE
UNIVERSITY PRESS

CAMBRIDGE
UNIVERSITY PRESS

University Printing House, Cambridge CB2 8BS, United Kingdom

One Liberty Plaza, 20th Floor, New York, NY 10006, USA

477 Williamstown Road, Port Melbourne, VIC 3207, Australia

314–321, 3rd Floor, Plot 3, Splendor Forum, Jasola District Centre, New Delhi – 110025, India

79 Anson Road, #06–04/06, Singapore 079906

Cambridge University Press is part of the University of Cambridge.

It furthers the University's mission by disseminating knowledge in the pursuit of education, learning and research at the highest international levels of excellence.

www.cambridge.org
Information on this title: www.cambridge.org/9781108464024

© Cambridge University Press 2018

First published 2018

20 19 18 17 16 15 14 13 12 11 10 9 8 7 6 5 4 3 2 1

Printed in Malaysia by Vivar Printing

A catalogue record for this publication is available from the British Library

ISBN 978-1-108-46402-4 Paperback

The questions, example answers, marks awarded and/or comments that appear in this book were written by the author(s). In examination, the way marks would be awarded to answers like these may be different.

This book has been compiled and authored by Jan Dangerfield, using questions from:

Cambridge International AS and A Level Mathematics: Mechanics 1 Coursebook (Revised edition) by Douglas Quadling and Julian Gilbey, that was originally published in 2016.

Cambridge International AS and A Level Mathematics: Mechanics 2 Coursebook by Douglas Quadling and Julian Gilbey, that was originally published in 2016.

A Level Mathematics for OCR A Student Book 1 (AS/Year 1) by Vesna Kadelburg and Ben Woolley

A Level Mathematics for OCR A Student Book 2 (Year 2) by Vesna Kadelburg and Ben Woolley

A Level Further Mathematics for AQA Mechanics Student Book 2 (AS/A Level) by Jess Barker, Nathan Barker, Michele Conway, Janet Such and Stephen Ward

Cover image: malerapaso/Getty Images

....

Contents

iii

Throughout this book you will notice particular features that are designed to help your learning. This section provides a brief overview of these features.

- Relate force to acceleration.
- Use combinations of forces to calculate their effect on an object.
- Include the force on an object due to gravity in a force diagram and calculations.
- Include the contact force on a force diagram and in calculations.

Learning objectives indicate the important concepts within each chapter and help you to navigate through the practice book.

WORKED EXAMPLE 2.1

A block of mass 16 kg is pushed across a smooth horizontal surface using a constant horizontal force of T newtons. The block starts from rest and takes 5 seconds to travel 20 metres. Find the value of T.

Answer

$s = 20$, $u = 0$, $t = 5$

$s = ut + \frac{1}{2}at^2$ so   Constant force so constant acceleration.

$a = 1.6 \, \mathrm{m \, s^{-2}}$

Newton's second law:

$T = 16 \times 1.6 = 25.6$ Resultant force = mass × acceleration.

END-OF-CHAPTER REVIEW EXERCISE 7

1 A raindrop of mass 0.005 kg has momentum of magnitude between 0.01 Ns and 0.02 Ns. Calculate the range of speeds of the raindrop.

2 A train of mass 300 000 kg is travelling along a straight track. The train speeds up from 4 m s⁻¹ to 6.5 m s⁻¹. Calculate the increase in the momentum of the train, measured in the direction of travel.

The **End-of-chapter review** contains exam-style questions covering all topics in the chapter. You can use this to check your understanding of the topics you have covered.

Extension material goes beyond the syllabus. It is highlighted by a red line to the left of the text.

TIP

The velocity jumps suddenly when $t = 5$. This can happen, for example, in a collision. The displacement still needs to be continuous.

Tip boxes contain helpful guidance about calculating or checking your answers.

Worked examples provide step-by-step approaches to answering questions. The left side shows a fully worked solution, while the right side contains a commentary explaining each step in the working.

Throughout each chapter there are multiple exercises containing practice questions. The questions are coded:

PS These questions focus on problem solving.

P These questions focus on proofs.

M These questions focus on modelling.

You should not use a calculator for these questions.

You can use a calculator for these questions.

Chapter 1
Velocity and acceleration

- Work with scalar and vector quantities for distance and speed.
- Use equations of constant acceleration.
- Sketch and read displacement–time graphs and velocity–time graphs.
- Solve problems with multiple stages of motion.

1.1 Displacement and velocity

WORKED EXAMPLE 1.1

A plane flies from Warsaw to Athens, a distance of 1600 km, at an average speed of 640 km h^{-1}. How long does the flight take?

Answer

$$\text{speed} = \frac{\text{distance}}{\text{time}}$$
State the equation to use.

$$\text{so time} = \frac{\text{distance}}{\text{speed}}$$
Rearrange the equation to make time the subject.

$$= \frac{1600}{640}$$
Use consistent units, substitute values into the equation.

$$= 2.5$$

Flight takes 2 hours 30 minutes.
Convert the decimal answer into hours and minutes.

EXERCISE 1A

1   How long will an athlete take to run 1500 metres at 7.5 m s^{-1}?

2   A train maintains a constant velocity of 60 m s^{-1} due south for 20 minutes. What is its displacement in that time? Give the distance in kilometres.

3   Some Antarctic explorers walking towards the South Pole expect to average 1.8 kilometres per hour. What is their expected displacement in a day in which they walk for 14 hours?

Questions 4 and 5 refer to the four points, A, B, C and D, which lie in a straight line with distances between them shown in the diagram. The displacement is measured from left to right.

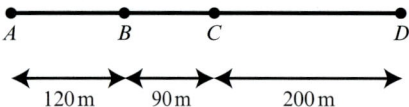

4 Find:

   a **i**   the displacement from D to A

       ii the displacement from D to B

   b **i**   the distance from D to B

       ii the distance from C to A

   c **i**   the total displacement when a particle travels from B to C and then to A

       ii the total displacement when a particle travels from C to D and then to A.

>
> **TIP**
>
> Remember displacement is a vector quantity, and distance is a scalar quantity.

PS **5** **a** **i** A particle travels from A to C in 23 seconds and then from C to B in 18 seconds. Find its average speed and average velocity.

       ii A particle travels from B to D in 38 seconds and then from D to A in 43 seconds. Find its average speed and average velocity.

   b **i**   A particle travels from B to D in 16 seconds and then back to B in 22 seconds. Find its average speed and average velocity.

       ii A particle travels from A to C in 26 seconds and then back to A in 18 seconds. Find its average speed and average velocity.

>
> **TIP**
>
> Remember speed is a scalar quantity, and velocity is a vector quantity.

6 Here is an extract from the diary of Samuel Pepys for 4 June 1666, written in London.

'We find the Duke at St James's, whither he is lately gone to lodge. So walking through the Parke we saw hundreds of people listening to hear the guns.'

These guns were at the battle of the English fleet against the Dutch off the Kent coast, a distance of between 110 and 120 km away. The speed of sound in air is $344\,\mathrm{m\,s^{-1}}$. How long did it take the sound of the gunfire to reach London?

7 Light travels at a speed of $3.00 \times 10^8\,\mathrm{m\,s^{-1}}$. Light from the star Sirius takes 8.65 years to reach the Earth. What is the distance of Sirius from the Earth in kilometres?

>
> **TIP**
>
> Consider how many seconds there are in 8.65 years.

1.2 Acceleration

WORKED EXAMPLE 1.2

A skateboarder travels down a hill in a straight line with constant acceleration. She starts with speed $1.5\,\mathrm{m\,s^{-1}}$ and finishes with speed $9.5\,\mathrm{m\,s^{-1}}$. The length of the hill is $22\,\mathrm{m}$.

a Find the time taken.

b Find the acceleration of the skateboarder.

Answer

a $s = 22, u = 1.5, v = 9.5$ .... Begin by listing the information given.

$s = \dfrac{1}{2}(u + v)t$ so $22 = \dfrac{1}{2}(1.5 + 9.5)t$ .... State the equation to be used and substitute in the known values.

$t = 4$ .... Rearrange the equation to find the time.

Time taken $= 4$ seconds .... Include the units in the final answer.

b $a = \dfrac{v - u}{t}$ .... Choose the equation to be used.

$= \dfrac{9.5 - 1.5}{4}$ .... Substitute in the known values.

$= 2$

Acceleration $= 2\,\mathrm{m\,s^{-2}}$ .... Include the units in the final answer.

EXERCISE 1B

1 Write the following quantities in the specified units, giving your answers to 3 significant figures.

TIP

Use velocities not speeds.

a i $3.6\,\mathrm{km\,h^{-1}}$ in $\mathrm{m\,s^{-1}}$ ii $62\,\mathrm{km\,h^{-1}}$ in $\mathrm{m\,s^{-1}}$

b i $5.2\,\mathrm{m\,s^{-1}}$ in $\mathrm{km\,h^{-1}}$ ii $0.26\,\mathrm{m\,s^{-1}}$ in $\mathrm{km\,h^{-1}}$

c i $120\,\mathrm{km\,h^{-2}}$ in $\mathrm{m\,s^{-2}}$ ii $450\,\mathrm{km\,h^{-2}}$ in $\mathrm{m\,s^{-2}}$

d i $0.82\,\mathrm{m\,s^{-2}}$ in $\mathrm{km\,h^{-2}}$ ii $2.7\,\mathrm{m\,s^{-2}}$ in $\mathrm{km\,h^{-2}}$

2 A police car accelerates from $15\,\mathrm{m\,s^{-1}}$ to $35\,\mathrm{m\,s^{-1}}$ in 5 seconds. The acceleration is constant. Illustrate this with a velocity–time graph. Use the equation $v = u + at$ to calculate the acceleration. Find also the distance travelled by the car in that time.

3

3   A marathon competitor running at $5\,\mathrm{m\,s^{-1}}$ puts on a sprint when she is 100 metres from the finish, and covers this distance in 16 seconds. Assuming that her acceleration is constant, use the equation $s = \dfrac{1}{2}(u + v)t$ to find how fast she is running as she crosses the finishing line.

4   Starting from rest, an aircraft accelerates to its take-off speed of $60\,\mathrm{m\,s^{-1}}$ in a distance of 900 metres. Assuming constant acceleration, find how long the take-off run lasts. Hence calculate the acceleration.

> **TIP**
>
> 'Rest' means not moving, so the velocity is zero.

5   A train is travelling at $80\,\mathrm{m\,s^{-1}}$ when the driver applies the brakes, producing a deceleration of $2\,\mathrm{m\,s^{-2}}$ for 30 seconds. How fast is the train then travelling, and how far does it travel while the brakes are on?

PS 6   A balloon at a height of 300 m is descending at $10\,\mathrm{m\,s^{-1}}$ and decelerating at a rate of $0.4\,\mathrm{m\,s^{-2}}$. How long will it take for the balloon to stop descending, and what will its height be then?

7   A train goes into a tunnel at $20\,\mathrm{m\,s^{-1}}$ and emerges from it at $55\,\mathrm{m\,s^{-1}}$. The tunnel is 1500 m long. Assuming constant acceleration, find how long the train is in the tunnel for, and the acceleration of the train.

PS 8   A cyclist riding at $5\,\mathrm{m\,s^{-1}}$ starts to accelerate, and 200 metres later she is riding at $7\,\mathrm{m\,s^{-1}}$. Find her acceleration, assumed constant.

1.3 Equations of constant acceleration

WORKED EXAMPLE 1.3

A train is travelling at $55\,\mathrm{m\,s^{-1}}$. The driver needs to reduce the speed to $35\,\mathrm{m\,s^{-1}}$ to pass through a junction. The deceleration must not exceed $0.6\,\mathrm{m\,s^{-2}}$. How far ahead of the junction should the driver begin to slow down the train?

Answer

Using the maximum deceleration:

$u = 55, v = 35, a = -0.6$ · · · · · · · · · · · · · · · · · · Begin by listing the information given. As we have deceleration, the acceleration is a negative value.

$v^2 = u^2 + 2as$ so $1225 = 3025 - 1.2s$ · · · · · · · · State the equation to be used and substitute in the known values.

$s = 1500$ .... Rearrange to find the distance.

The driver should start to slow down at least .... Include the units in the final answer
1500 m ahead of the junction. and clarify the answer in the context
of the question.

EXERCISE 1C

TIP

Decide which
variable to eliminate.

P 1 Use the formulae $v = u + at$ and $s = \frac{1}{2}(u + v)t$ to prove that
$s = ut + \frac{1}{2}at^2$.

P 2 **a** Use the formulae $s = ut + \frac{1}{2}at^2$ and $v = u + at$ to derive the
formula $s = vt - \frac{1}{2}at^2$.

b A particle moves with constant acceleration $3.1\,\mathrm{m\,s^{-2}}$. It travels
300 m in the first 8 seconds. Find its speed at the end of the
8 seconds.

P 3 Use the formulae $v = u + at$ and $s = ut + \frac{1}{2}at^2$ to derive the
formula $v^2 = u^2 + 2as$.

4 An ocean liner leaves the harbour entrance travelling at $3\,\mathrm{m\,s^{-1}}$, and
accelerates at $0.04\,\mathrm{m\,s^{-2}}$ until it reaches its cruising speed of $15\,\mathrm{m\,s^{-1}}$.

a How far does it travel in accelerating to its cruising speed?

b How long does it take to travel 2 km from the harbour entrance?

5 A boy kicks a football up a slope with a speed of $6\,\mathrm{m\,s^{-1}}$. The ball
decelerates at $0.3\,\mathrm{m\,s^{-2}}$. How far up the slope does it roll?

6 A cyclist comes to the top of a hill 165 metres long travelling at $5\,\mathrm{m\,s^{-1}}$,
and free-wheels down it with an acceleration of $0.8\,\mathrm{m\,s^{-2}}$. Write
expressions for his speed and the distance he has travelled after
t seconds. Hence find how long he takes to reach the bottom of the
hill, and how fast he is then travelling.

PS 7 A particle reduces its speed from $20\,\mathrm{m\,s^{-1}}$ to $8.2\,\mathrm{m\,s^{-1}}$ while travelling
100 m. Assuming it continues to move with the same constant
acceleration, how long will it take to travel another 20 m?

PS 8 A particle moves with constant deceleration of $3.6\,\mathrm{m\,s^{-2}}$. It travels
350 m while its speed halves. Find the time it takes to do this.

9 A car reduces its speed from $18\,\mathrm{m\,s^{-1}}$ to $9\,\mathrm{m\,s^{-1}}$ while travelling
200 m. Assuming the car continues to move with the same uniform
acceleration, how much further will it travel before it stops?

10 a A particle moves in a straight line with constant acceleration $a = -3.4\,\mathrm{m\,s^{-2}}$. At $t = 0$ its velocity is $u = 6\,\mathrm{m\,s^{-1}}$. Find its maximum displacement from the starting point.

b Explain why this is not the maximum distance from the starting point.

M **11** A car travelling at $10\,\mathrm{m\,s^{-1}}$ is 25 metres from a pedestrian crossing when the traffic light changes from green to amber. The light remains at amber for 2 seconds before it changes to red. The driver has two choices: to accelerate so as to reach the crossing before the light changes to red, or to try to stop at the light. What is the least acceleration which would be necessary in the first case, and the least deceleration which would be necessary in the second?

M **12** A cheetah is pursuing an impala. The impala is running in a straight line at a constant speed of $16\,\mathrm{m\,s^{-1}}$. The cheetah is $10\,\mathrm{m}$ behind the impala, running at $20\,\mathrm{m\,s^{-1}}$ but tiring, so that it is decelerating at $1\,\mathrm{m\,s^{-2}}$. Find an expression for the gap between the cheetah and the impala t seconds later. Will the impala get away?

1.4 Displacement–time graphs and multi-stage problems

WORKED EXAMPLE 1.4

The diagram shows the displacement–time graph for a cyclist moving in a straight line.

a Find the velocity of the cyclist over the first 50 seconds.

b Estimate the times when the velocity of the cyclist is $0\,\mathrm{m\,s^{-1}}$.

c Find the greatest (positive) velocity over the 100 seconds.

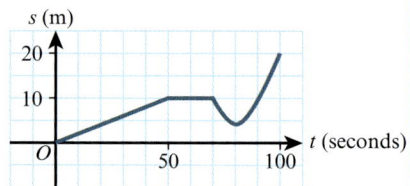

Answer

a $\dfrac{10}{50} = 0.2\,\mathrm{m\,s^{-1}}$ .... The gradient of a displacement–time graph is equal to the velocity.

b $t = 50$ to 70 and at approximately 82 seconds .... As the gradient represents the velocity, we look at when the gradient is zero.

c Velocity = gradient of graph

Greatest (positive) velocity is $t = 90$ to 100 .... This is where the graph is steepest.

In this time s increases from 10 to 20

$\text{Velocity} = \dfrac{20 - 10}{100 - 90} = 1\,\mathrm{m\,s^{-1}}$ .... Gradient = increase in height/change in horizontal distance.

EXERCISE 1D

1   A particle moves in a straight line. Its displacement from point P is shown on the displacement–time graph.

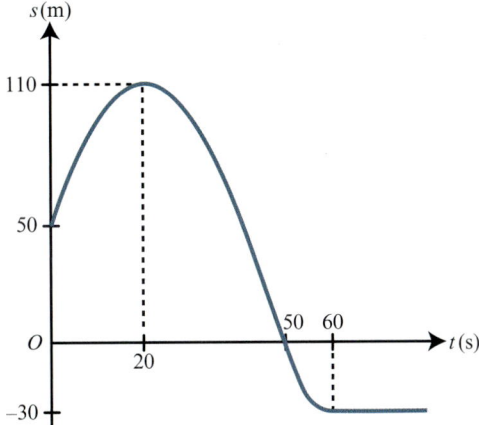

a   How far from P does the particle start?

b   In the first 20 seconds, is the particle moving towards P or away from it?

c   What happens when $t = 20$ seconds?

d   What happens after 60 seconds?

e   At what time does the particle pass P?

f   Is the particle's speed increasing or decreasing during the first 20 seconds?

g   Is the particle's speed increasing or decreasing between 50 and 60 seconds? What about its velocity?

h   Find the total distance travelled by the particle in the first 60 seconds.

PS   2   For each displacement–time graph, draw the corresponding straight line velocity–time graph:

a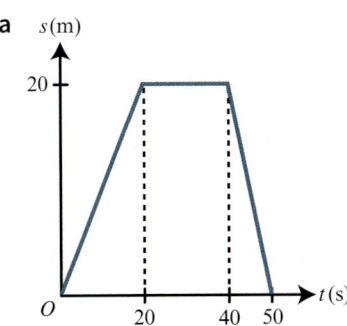

b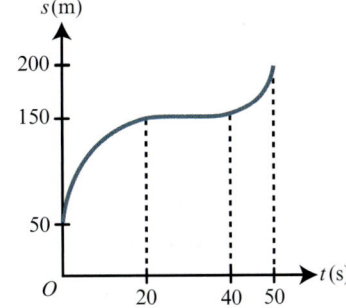

c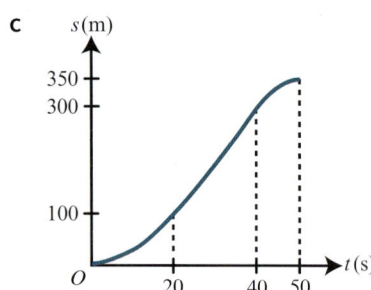

d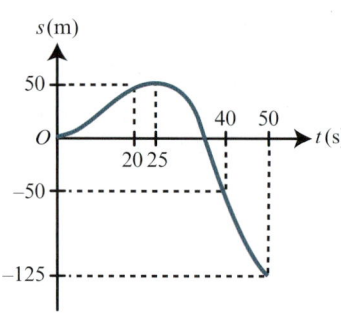

3 This displacement–time graph represents the motion of a particle moving in a straight line. The particle passes point A when $t = 0$.

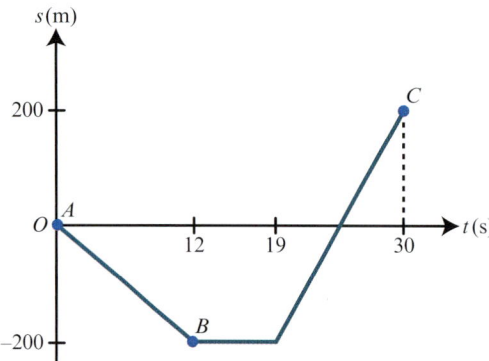

The particle is at point B when $t = 12$ and at point C when $t = 30$.

a Describe what happens between $t = 12$ and $t = 19$.

b Write down the displacement of C from A. Hence find the average velocity of the particle during the 30 seconds.

c Find the average speed of the particle during the 30 seconds.

 4 A cyclist is free-wheeling down a long straight hill. The times between passing successive kilometre posts are 100 seconds and 80 seconds. Assuming his acceleration is constant, find this acceleration.

5 A train is slowing down with constant deceleration. It passes a signal at A, and after successive intervals of 40 seconds it passes points B and C, where $AB = 1800\,\text{m}$ and $BC = 1400\,\text{m}$.

a How fast is the train moving when it passes A?

b How far from A does it come to a stop?

 6 A particle is moving along a straight line with constant acceleration. In an interval of T seconds it moves D metres; in the next interval of $3T$ seconds it moves $9D$ metres.

How far does it move in a further interval of T seconds?

 TIP

Each individual km can be considered, or the combined motion for 2 km can be considered.

8

7 A ball is thrown vertically upwards, travels up and then down; when the ball hits the ground it bounces up again.

The following graph shows the height of the ball above the ground against time.

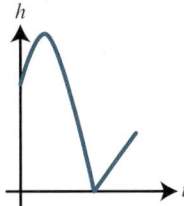

Sketch a graph to show the distance that the ball has travelled against time.

8 Amir and Sofia start side by side on the starting line of a 100 m track. Amir runs the 100 metres at a constant speed of v m s^{-1}. Sofia starts from rest 1 second after Amir and accelerates at a constant 0.5 m s^{-2}.

    a When $v = 4$, how far has Amir run when Sofia overtakes him?

    b What happens when $v = 5$?

9 A particle starts from rest and moves with constant acceleration.

    a Sketch the displacement–time graph.

In the first 4 seconds the particle moves 16 metres.

    b Find how far the particle travels in the next 4 seconds.

10 A motorbike and a car are waiting side by side at traffic lights. When the lights turn to green, the motorbike accelerates at 2.5 m s^{-2} up to a top speed of 20 m s^{-1}, and the car accelerates at 1.5 m s^{-2} up to a top speed of 30 m s^{-1}. Both then continue to move at constant speed.

    a Using the same axes, sketch the displacement–time graphs.

    b After what time will the motorbike and the car again be side by side?

    c What is the greatest distance that the motorbike is in front of the car?

11 The displacement, in metres, of a particle is plotted against time, in seconds. The resulting displacement–time graph is modelled as a quadratic equation that passes through $(0, 20)$, $(10, 5)$ and $(12, 20)$.

At some time the displacement is x metres and 4 seconds later the displacement is again x metres. Find the value of x.

1.5 Velocity–time graphs and multi-stage problems

WORKED EXAMPLE 1.5

A train moves along a straight track. The train passes a signal at time $t = 0$ seconds and is moving with a velocity of $18 \, \text{m s}^{-1}$. The train accelerates at $2 \, \text{m s}^{-2}$ for 5 seconds, travels at a constant velocity for some time, and then decelerates at $1.4 \, \text{m s}^{-2}$ until it stops at a station. The signal is 1000 m from the station. Find the time at which the train reaches the station.

Answer

When $t = 0$, $v = 18$ .... State the initial values.

When $t = 5$, $v = 18 + 2 \times 5 = 28$ .... Use $v = u + at$ to find the velocity after 5 seconds.

The velocity–time graph looks like this: .... Although we don't have all the information for a velocity–time graph, drawing a sketch of what we do know may help to answer the question.

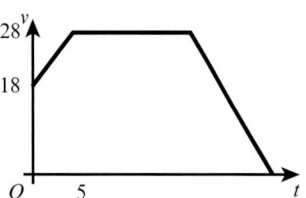

While accelerating, the train travels .... Use $s = \dfrac{1}{2}(u + v)t$ to find the distance
$0.5(18 + 28) \times 5 = 115$ metres travelled during the first section of the motion.

The deceleration phase takes .... Use $\dfrac{v - u}{a} = t$ to find the time taken to
$\dfrac{0 - 28}{-1.4} = 20$ seconds deceleration to rest.

The distance travelled while decelerating is ..... Use $s = \dfrac{1}{2}(u + v)t$ to find the distance
$0.5(28 + 0) \times 20 = 280$ metres travelled during the final section of the motion.

So the distance travelled at $28 \, \text{m s}^{-1}$ is .... Use the total distance to find the distance
$1000 - 115 - 280 = 605$ metres travelled during the constant speed section of motion.

This takes $605 \div 28 = 21.6$ seconds .... Use time $= \dfrac{\text{distance}}{\text{speed}}$.

$5 + 21.6 + 20 = 46.6$ .... Finally add the time for each of the three sections.

The train reaches the station at time
$t = 46.6$ seconds

1 For each velocity–time graph, find:

a the acceleration from A to B and from C to D

b the total distance travelled.

i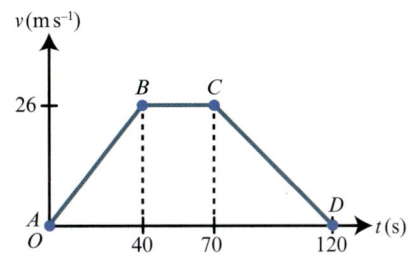

ii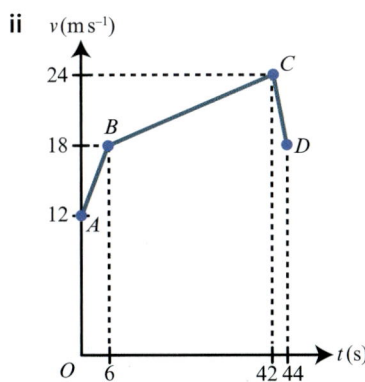

iii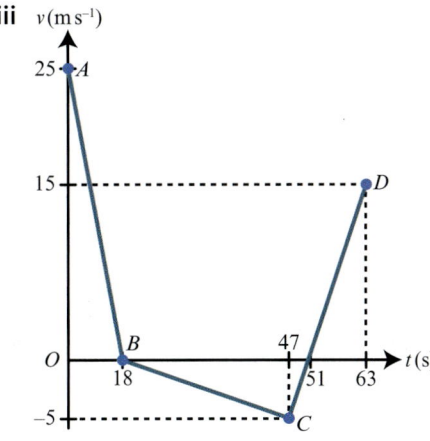

2 For each of these descriptions of motion, draw the velocity–time graph and find the total distance travelled.

a A particle accelerates uniformly from $20\,\text{m s}^{-1}$ to $32\,\text{m s}^{-1}$ in 15 seconds, then moves with constant speed for 25 seconds and finally decelerates uniformly and comes to rest in another 10 seconds.

b An object starts from rest and accelerates at $2.5\,\mathrm{m\,s^{-2}}$ for 12 seconds. It then moves with a constant velocity for 8 seconds and finally decelerates at $6\,\mathrm{m\,s^{-2}}$ until it comes to rest.

c A particle accelerates uniformly from $11\,\mathrm{m\,s^{-1}}$ to $26\,\mathrm{m\,s^{-1}}$ with acceleration $0.4\,\mathrm{m\,s^{-2}}$. It then decelerates at $2\,\mathrm{m\,s^{-2}}$ until it comes to rest.

3 A train travelling at $20\,\mathrm{m\,s^{-1}}$ starts to accelerate with constant acceleration. It covers the next kilometre in 25 seconds. Use the equation $s = ut + \frac{1}{2}at^2$ to calculate the acceleration. Find also how fast the train is moving at the end of this time. Illustrate the motion of the train with a velocity–time graph. How long does the train take to cover the first half kilometre?

4 A long-jumper takes a run of 30 metres to accelerate to a speed of $10\,\mathrm{m\,s^{-1}}$ from a standing start. Find the time he takes to reach this speed, and hence calculate his acceleration. Illustrate his run-up with a velocity–time graph.

PS **5** A particle moves in a straight line, starting from rest at point P. It accelerates for 5 seconds, until it reaches a speed of $16\,\mathrm{m\,s^{-1}}$. It maintains this speed for T seconds and then decelerates at $2\,\mathrm{m\,s^{-2}}$ until it comes to rest at point Q.

a Sketch the velocity–time graph to represent the motion of the particle.

b Given the average speed of the particle on the journey from P to Q is $12\,\mathrm{m\,s^{-1}}$, find the value of T.

6 A cyclist travels from A to B, a distance of 240 metres. He passes A at $12\,\mathrm{m\,s^{-1}}$, maintains this speed for as long as he can, and then brakes so that he comes to a stop at B. If the maximum deceleration he can achieve when braking is $3\,\mathrm{m\,s^{-2}}$, what is the least time in which he can get from A to B?

PS **7** Two villages are 900 metres apart. A car leaves the first village travelling at $15\,\mathrm{m\,s^{-1}}$ and accelerates at $\frac{1}{2}\,\mathrm{m\,s^{-2}}$ for 30 seconds. How fast is it then travelling, and what distance has it covered in this time?

The driver now sees the next village ahead, and decelerates so as to enter it at $15\,\mathrm{m\,s^{-1}}$. What constant deceleration is needed to achieve this? How much time does the driver save by accelerating and decelerating, rather than covering the whole distance at $15\,\mathrm{m\,s^{-1}}$?

M 8 A cyclist starts at the bottom of a hill moving at a speed of $13.5\,\mathrm{m\,s^{-1}}$. She moves with a constant deceleration of $0.9\,\mathrm{m\,s^{-2}}$, reaching the top of the hill 9.2 seconds later. She then accelerates down the hill at $1.6\,\mathrm{m\,s^{-2}}$ for 86 m. Find the speed of the cyclist when she reaches the bottom of the hill.

9 A car comes to a stop from a speed of $30\,\mathrm{m\,s^{-1}}$ in a distance of 804 m. The driver brakes so as to produce a deceleration of $\frac{1}{2}\,\mathrm{m\,s^{-2}}$ to begin with, and then brakes harder to produce a deceleration of $\frac{3}{2}\,\mathrm{m\,s^{-2}}$. Find the speed of the car at the instant when the deceleration is increased, and the total time the car takes to stop.

10 A ball is dropped from a height of 2.6 m above the surface of a water well and falls freely under gravity. After it enters the water, the ball's acceleration decreases to $1.2\,\mathrm{m\,s^{-2}}$. It reaches the bottom of the well 0.9 seconds after reaching the surface of the water. Assuming the acceleration through the water is constant over a short period of time, find the depth of the water in the well.

 TIP

Use a value of $10\,\mathrm{m\,s^{-2}}$ for the acceleration whilst the ball is above the surface of the water.

M 11 A roller-skater increases speed from $4\,\mathrm{m\,s^{-1}}$ to $10\,\mathrm{m\,s^{-1}}$ in 10 seconds at a constant rate.

   a What is her average velocity over this period?

   b For what proportion of the time is she moving at less than her average velocity?

   c For what proportion of the distance is she moving at less than her average velocity?

12 A car starts from rest at time $t = 0$. It accelerates uniformly until its speed reaches $V\,\mathrm{m\,s^{-1}}$. It travels at constant speed for 12 seconds and then decelerates uniformly, coming to rest when $t = 26$. The total distance travelled by the car is 840 m. Find the value of V.

1.6 Graphs with discontinuities

WORKED EXAMPLE 1.6

In a game of miniature golf, a player hits the ball from the tee so that it rolls along the ground with an initial speed of $10\,\mathrm{m\,s^{-1}}$. The ball is subject to a deceleration of constant magnitude $D\,\mathrm{m\,s^{-2}}$ throughout its motion.

The ball hits a small wall 5 metres from the tee and returns past the tee. It hits another wall 1 metre beyond the tee and returns to the tee where it stops, having been rolling for T seconds. The ball loses 50% of its speed each time it hits a wall.

   a Sketch the velocity–time graph for the motion.

   b Find the value of D.   c Find the value of T.

Answer

a   Taking the direction from the tee to
    the first wall as the positive direction:

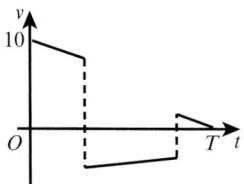

Each time the ball hits a wall the
velocity changes sign.

b   From the tee to the first wall:

$s = 5, u = 10, a = -D$

Begin by listing the information given,
noting the acceleration is negative as
it is a deceleration.

Using $v^2 = u^2 + 2as$:

Choose the most suitable equation.

$v^2 = 100 - 10D$

Substitute in the known values.

Ball hits the wall with velocity
$\sqrt{100 - 10D}$ and leaves with velocity
$-0.5\sqrt{100 - 10D}$.

The ball leaves the wall with half the
previous velocity and the velocity
changes sign.

From the first wall to the second wall:

$s = -6, u = -0.5\sqrt{100 - 10D}, a = D$

List the new information known.

Using $v^2 = u^2 + 2as$:

Choose the most suitable equation.

$v^2 = 0.25(100 - 10D) - 12D = 25 - 14.5D$

Substitute in the known values and
simplify the answer.

Ball hits the second wall with velocity
$-\sqrt{25 - 14.5D}$ and leaves with velocity
$0.5\sqrt{25 - 14.5D}$.

Again the ball leaves the wall with half
the previous velocity and the velocity
changes sign.

From the second wall to the tee:

$s = 1, u = 0.5\sqrt{25 - 14.5D}, v = 0, a = -D$

List the new information known.

Using $v^2 = u^2 + 2as$:

Choose the most suitable equation.

$0 = 0.25(25 - 14.5D) - 2D$

Substitute in the known values.

$D = \dfrac{10}{9} = 1.11$

Solve to find D.

c   From the tee to the first wall:

$s = 5, u = 10, v = 9.43, a = -1.11$

List the new information known.

Using $v = u + at$:

Choose the most suitable equation.

$9.43 = 10 - 1.11t$ so $t = 0.514$

Substitute in the known values and find
the time.

14

From the first wall to the second wall:

$s = -6, u = -4.72, v = -2.98, a = 1.11$ · · · · List the new information known.

Using $v = u + at$: · · · · · · · · · · · · · · · Choose the most suitable equation.

$-2.98 = -4.72 + 1.11t$ so $t = 1.559$ · · · · Substitute in the known values and find the time.

From the second wall to the tee:

$s = 1, u = 1.49, v = 0, a = -1.11$ · · · · · · List the new information known.

Using $v = u + at$: · · · · · · · · · · · · · · · Choose the most suitable equation.

$0 = 1.49 - 1.11t$ so $t = 1.342$ · · · · · · · Substitute in the known values and find the time.

$T = 0.518 + 1.559 + 1.342 = 3.42 \text{ seconds}$ · · · Add together the times for each of the three sections of motion.

EXERCISE 1F

M **1** An ice hockey puck slides across the ice with a constant speed of 15 m s^{-1}. The puck travels 10 metres, bounces off the board at the edge of the ice, losing 20% of its speed, and returns with constant velocity to the point where it was hit.

How long does the player who hit the puck have to get out of the way, to avoid being hit by the puck?

PS **2** A car is travelling at 20 m s^{-1} when the driver brakes heavily. The car is instantly subject to a constant deceleration that lasts for 1.5 seconds and slows the speed of the car to 5 m s^{-1}. The driver then eases off the brakes so the deceleration changes to a different constant value that brings the car to rest after travelling a further 50 metres.

Find:

a the total time taken

b the total distance travelled from when the brakes are applied to when the car stops.

PS **3** A ball is dropped from a window that is 10 m above the ground. The initial speed of the ball is 0 m s^{-1} and it accelerates at 9.8 m s^{-2} while it is falling. The ball bounces and loses 10% of its speed in the bounce.

a How high does the ball rise after the first bounce?

b How long does it take until the ball hits the ground for the second time?

A person spots the bouncing ball and catches it as it approaches the ground for the first time that is more than 10 seconds from when the ball first bounced.

c How many more bounces does the ball make after the first bounce before it is caught?

END-OF-CHAPTER REVIEW EXERCISE 1

> **TIP**
>
> Graphs and equations of motion may both be helpful.

1 A train leaves a station, starting from rest, with a constant acceleration of $a\,\mathrm{m\,s^{-2}}$. It reaches a signal 100 seconds later at a speed of $40\,\mathrm{m\,s^{-1}}$. Find:

  a the value of a

  b the distance between the station and the signal.

 2 A woman skis down a slope with constant acceleration. She starts from rest and is travelling at $25\,\mathrm{m\,s^{-1}}$ when she reaches the bottom of the slope. The slope is 125 m long. Find:

  a her acceleration down the slope

  b the time taken to reach the bottom of the slope.

 3 A particle moves along a straight line ABC with constant acceleration. $AB = 50\,\mathrm{cm}$ and $BC = 150\,\mathrm{cm}$. After passing through A, the particle travels for 2 seconds before passing through B, and for a further 3 seconds before passing through C. Find the acceleration of the particle and the speed with which it reaches C.

> **TIP**
>
> Remember to use SI units throughout.

4 A car is travelling at $V\,\mathrm{m\,s^{-1}}$ along a straight road and passes point A at $t = 0$. When the car is at point A the driver sees a pedestrian crossing the road at a point B ahead and decelerates at $1\,\mathrm{m\,s^{-2}}$ for 6 seconds. The car then travels at a constant speed and reaches B after a further 6 seconds. The distance AB is at 180 m.

  a Sketch a velocity–time graph for the car's journey.

  b Determine the value of V.

5 A car is travelling along a road. It passes point A at a constant speed of $V\,\mathrm{m\,s^{-1}}$, and drives for T seconds at this speed. It then accelerates at a constant rate for 6 seconds until it reaches a speed of $2V\,\mathrm{m\,s^{-1}}$. Maintaining this speed, it arrives at point B after a further $2T$ seconds. The total distance travelled between A and B was 528 m and the average speed was $20\,\mathrm{m\,s^{-1}}$. Find V and T.

6 A goods train travels along a straight track between two signals. The train has speed $3\,\mathrm{m\,s^{-1}}$ as it passes through the first signal. The train moves with constant acceleration $0.5\,\mathrm{m\,s^{-2}}$ for the first 30 s, then with constant speed for 180 s, and finally with constant deceleration $0.12\,\mathrm{m\,s^{-2}}$ to come to rest at the second signal.

a Find the total time taken for the train to travel between the two signals.

b Calculate the distance between the two signals.

c Find, to the nearest whole number, the percentage of the distance between the two signals that the train travels at more than half the maximum speed.

(M) 7 Two runners, Ayesha and Fatima, are completing a long-distance race. They are both running at $5\,\mathrm{m\,s^{-1}}$, with Ayesha 10 m behind Fatima. When Fatima is 50 m from the finish line, Ayesha accelerates but Fatima doesn't. What is the least acceleration Ayesha must produce to overtake Fatima?

If instead Fatima accelerates at $0.1\,\mathrm{m\,s^{-2}}$ up to the finish line, what is the least acceleration Ayesha must produce?

(M) 8 A woman stands on the bank of a frozen lake with a dog by her side. She slides a bone across the ice at a speed of $3\,\mathrm{m\,s^{-1}}$ and at the same instant the dog begins to chase the bone. The bone slows down with deceleration $0.4\,\mathrm{m\,s^{-2}}$, and the dog chases it with acceleration $0.6\,\mathrm{m\,s^{-2}}$. How far out from the bank does the dog catch up with the bone?

9 A man is running for a bus at $3\,\mathrm{m\,s^{-1}}$. When he is 100 m from the bus stop, the bus passes him going at $8\,\mathrm{m\,s^{-1}}$. If the deceleration of the bus is constant, at what constant rate should the man accelerate so as to arrive at the bus stop at the same instant as the bus?

10 The distance from A to B is 10 000 m. A car starts from rest at A and accelerates uniformly until it has travelled 2000 m. The car travels at a constant speed $V\,\mathrm{m\,s^{-1}}$ for 300 s and then decelerates uniformly for T s to come to rest at B.

The average speed of the car between A and B is $20\,\mathrm{m\,s^{-1}}$.

a Find the value of V.

b Find the value of T.

(PS) 11 If a ball is placed on a straight sloping track and then released from rest, the distances that it moves in successive equal intervals of time are found to be in the ratio $1:3:5:7:\ldots$. Show that this is consistent with the theory that the ball rolls down the track with constant acceleration.

Chapter 2
Force and motion in one dimension

- Relate force to acceleration.
- Use combinations of forces to calculate their effect on an object.
- Include the force on an object due to gravity in a force diagram and calculations.
- Include the normal contact force on a force diagram and in calculations.

 TIP

Unless otherwise stated, g should be taken as $10\,\mathrm{m\,s^{-2}}$.

2.1 Newton's first law and relation between force and acceleration

WORKED EXAMPLE 2.1

A block of mass 16 kg is pushed across a smooth horizontal surface using a constant horizontal force of T newtons. The block starts from rest and takes 5 seconds to travel 20 metres. Find the value of T.

Answer

$s = 20,\ u = 0,\ t = 5$

$s = ut + \dfrac{1}{2}at^2$ so   Constant force so constant acceleration.

$a = 1.6\,\mathrm{m\,s^{-2}}$

Newton's second law:

$T = 16 \times 1.6 = 25.6$ Resultant force = mass × acceleration.

 TIP

We often need to use constant acceleration equations of motion before we can use Newton's second law.

EXERCISE 2A

1   Find the magnitude of the force, in newtons, acting on the object in each case.

   a   A crate of mass 53 kg moves with constant acceleration of $2.6\,\mathrm{m\,s^{-2}}$.

   b   A stone of mass 1.5 kg is pushed across ice and decelerates at a constant rate of $0.3\,\mathrm{m\,s^{-2}}$.

   c   A truck of mass 6 tonnes accelerates uniformly at $1.2\,\mathrm{m\,s^{-2}}$.

   d   A toy car of mass 230 g moves with constant acceleration of $3.6\,\mathrm{m\,s^{-2}}$.

   e   A box of mass 32 kg is dragged across the floor in a straight line, at a constant speed of $5.2\,\mathrm{m\,s^{-1}}$.

   f   A ball of mass 120 g falls with a constant acceleration of $9.8\,\mathrm{m\,s^{-2}}$.

   g   A book of mass 340 g rests on a horizontal table.

2   The engine of a car of mass 800 kg which is travelling along a straight horizontal road, is producing a driving force of 1200 N. Assuming that there are no forces resisting the motion, calculate the acceleration of the car.

3   A van is pulling a broken-down car of mass 1200 kg along a straight horizontal road. The only force acting on the car which affects the motion of the car is the tension in the horizontal towbar. Calculate the acceleration of the car when the tension is 750 N.

4   A stone of mass 120 g is pushed across ice with a speed of $3.2\,\mathrm{m\,s^{-1}}$. It comes to rest 8 seconds later. Find the magnitude of the frictional force acting on the stone.

5   A crate of mass 28 kg is pulled across a horizontal floor. The pulling force acting on the crate is 260 N. Assuming that any frictional forces can be ignored, how long does it take for the crate to accelerate from rest to $2.5\,\mathrm{m\,s^{-1}}$?

6   Two children are sliding a box from one to the other on a frozen lake. The box, of mass 0.4 kg, leaves one with speed $5\,\mathrm{m\,s^{-1}}$ and reaches the other, who is 8 m away, after 2.5 s. Calculate the deceleration of the box, and find the frictional force resisting the motion of the box.

7   A hockey player hits a stationary ball of mass 0.2 kg. The contact time between the stick and the ball is 0.15 s and the force exerted on the ball by the stick is 60 N. Find the speed with which the ball leaves the stick.

8   A car of mass 1000 kg runs out of petrol and comes to rest just 30 m from a garage. The car is pushed, with a force of 120 N, along the horizontal road towards the garage. Calculate the acceleration of the car and find the time it takes to reach the garage.

9   A van of mass 2.3 tonnes, travelling in a straight line, decelerates under the action of a constant braking force. Its speed decreases from $50\,\mathrm{km\,h^{-1}}$ to $30\,\mathrm{km\,h^{-1}}$ while it covers the distance of 650 m. Find the magnitude of the braking force.

10 A girl pulls a toy truck with a constant horizontal force of 23 N. The truck starts from rest and accelerates uniformly, travelling 16 m in 3 seconds. Find the mass of the truck.

2.2 Combinations of forces

WORKED EXAMPLE 2.2

A motor boat of mass 1500 kg is propelled forwards by a driving force of 280 N from the engine. A constant resistance force of 200 N acts on the boat. Find the acceleration of the boat.

Answer

Newton's second law: Draw a force diagram.

200 N ← □ → 280 N

TIP

Always draw a force diagram to ensure you have considered all forces in the problem and added them into the relevant equations.

$280 - 200 = 1500a$ ········· Resultant force = mass × acceleration.

$a = 0.0533 \, \text{m s}^{-2}$ ········· Include units in the final answer.

EXERCISE 2B

1 Two people attempt to push-start a car on a horizontal road. One person pushes with a force of 100 N; the other with a force of 80 N. The car starts to accelerate constantly at $0.15 \, \text{m s}^{-2}$. Assuming these are the only horizontal forces acting, find the mass of the car.

2 A sledge of mass m kg is pushed horizontally through the snow by a force of 40 N. There is resistance to its motion of magnitude 10 N as shown in the diagram. If the sledge is accelerating at $1.5 \, \text{m s}^{-2}$, find m.

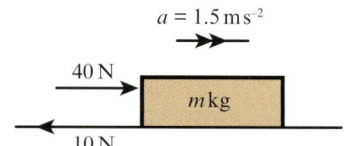

3 A motorcyclist moves with an acceleration of $5 \, \text{m s}^{-2}$ along a horizontal road against a total resistance of 120 N. The total mass of the rider and the motorcycle is 400 kg. Find the driving force provided by the engine.

4 Three men are trying to move a large rubbish bin. Two of the men are pushing horizontally with forces of magnitude 120 N and 150 N and one man is pulling with a horizontal force of magnitude X N. The frictional force resisting the motion is 385 N. Given that the bin does not move, find the value of X.

5 A car of mass 1200 kg is moving with a constant speed of $20 \, \text{m s}^{-1}$ in a horizontal straight line, against a resisting force of 300 N. What driving force is being provided to sustain this motion?

The driver speeds up uniformly over the next 30 s to reach a speed of $30 \, \text{m s}^{-1}$. Assuming that the resisting force remains at 300 N, calculate the extra driving force produced.

6 A student is dragging a luggage trunk of mass 85 kg along a corridor with an acceleration of $0.18 \, \text{m s}^{-2}$. The horizontal force the student exerts is 180 N. Find the frictional force between the floor and the trunk.

7 A porter is pushing a heavy crate of mass M kg along a horizontal floor with a horizontal force of 180 N. The resistance to motion has magnitude $3M$ newtons. Given that the acceleration of the crate is $0.45 \, \text{m s}^{-2}$, find the value of M.

8 A motor-boat of mass 8 tonnes is travelling along a straight course with a constant speed of $28 \, \text{km h}^{-1}$. The constant force driving the boat forward has magnitude 780 N. Find the force resisting motion, assumed constant.

The engine is now shut off. Calculate, to the nearest second, the time it takes the motor-boat to stop, assuming that the resistance remains the same as before.

9 A child pushes a box of mass 8 kg horizontally with a constant force of 28 N. The frictional force between the box and the floor is 12 N. Find the acceleration of the box.

10 Two men are pushing a car, in a straight line each using an equal force of magnitude F N. The resistance to motion has magnitude 420 N. The mass of the car is 850 kg and it is moving at a constant speed of 6 km h^{-1}. Find the value of F.

2.3 Weight and motion due to gravity

WORKED EXAMPLE 2.3

A ball is thrown vertically upwards from ground level with initial speed 12 m s^{-1}. At the same instant as the ball is released a second ball is dropped from rest from a window. The two balls collide after one second. Calculate the height of the window.

Answer

For the first ball:

$u = 12$, $a = -10$, $t = 1$ Use upwards as the positive direction, so acceleration is negative.

$s = ut + \dfrac{1}{2}at^2$ so $s = 7$ $12 \times 1 + 0.5(-10 \times 1^2) = 12 - 5 = 7$

The ball reaches a height of 7 m after 1 second.

For the second ball:

$u = 0$, $a = 10$, $t = 1$ Use downwards as the positive direction, so acceleration is positive.

$s = ut + \dfrac{1}{2}at^2$ so $s = 5$ $0 + 0.5(10 \times 1^2) = 5$

The ball falls 5 m in 1 second.

$7 + 5 = 12$ Add the distance the first ball rises to the distance the second ball falls.

The height of the window is 12 m.

EXERCISE 2C

1 A piece of luggage weighs 170 N. Find its mass.

2 A lift bringing miners to the surface of a mine shaft is moving with an acceleration of 1.2 m s^{-2}. The total mass of the cage and the miners is 1600 kg. Find the tension in the lift cable.

M 3 The total mass of a hot-air balloon, occupants and ballast, is 1300 kg. What is the upthrust on the balloon when it is travelling vertically upwards with constant velocity?

The occupants now release 50 kg of ballast. Assuming no air resistance, find the immediate acceleration of the balloon. (The upthrust is the upward buoyancy force, which does not change when the ballast is thrown out.)

4   A crate is lowered from a window of a space ship on Mars, using a rope. The tension in the rope is 328 N and the crate is descending at a constant speed. Given that the gravitational acceleration on Mars is $3.7\,\text{m}\,\text{s}^{-2}$, find the mass of the crate.

5   A boy of mass 45 kg is stranded on a beach as the tide comes in. A rescuer of mass 75 kg is lowered down, by rope, from the top of the cliff. The boy and rescuer are raised together, initially with a constant acceleration of $0.6\,\text{m}\,\text{s}^{-2}$. Find the tension in the rope for this stage of the ascent.

As they near the top of the cliff, the tension in the rope is 1020 N and they are moving with a constant deceleration. Calculate the magnitude of this deceleration.

6   The tension in the vertical cable of a crane is 1250 N when it is raising a girder with constant speed. Calculate the tension in the cable when it is raising the girder with an acceleration of $0.2\,\text{m}\,\text{s}^{-2}$, assuming no air resistance.

PS   7   A load of mass M is raised with constant acceleration, from rest, by a rope. The load reaches a speed of v in a distance of s. The tension in the rope is T. Find an expression for s in terms of M, T, v and g.

PS   8   A load of weight 7 kN is being raised from rest with constant acceleration by a cable. After the load has been raised 20 metres, the cable suddenly becomes slack. The load continues upwards for a distance of 4 metres before coming to instantaneous rest. Assuming no air resistance, find the tension in the cable before it became slack.

2.4 Normal contact force and motion in a vertical line

WORKED EXAMPLE 2.4

Two boxes are stacked one on top of another. Each box has mass 15 kg. The lower box is raised upwards at a constant $2\,\text{m}\,\text{s}^{-2}$. Calculate the magnitude of the contact force between the boxes.

Answer

The forces acting on the upper box are:

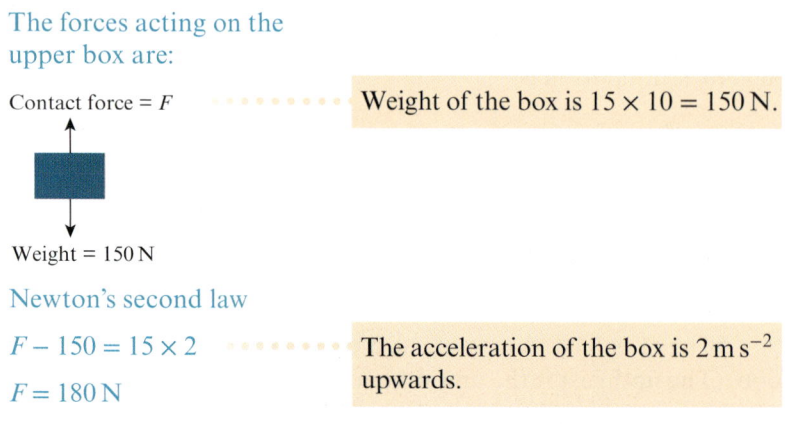

Contact force $= F$ ········· Weight of the box is $15 \times 10 = 150\,\text{N}$.

Weight $= 150\,\text{N}$

Newton's second law

$F - 150 = 15 \times 2$ ········· The acceleration of the box is $2\,\text{m}\,\text{s}^{-2}$ upwards.

$F = 180\,\text{N}$

 TIP

Do not include units on diagrams where forces are indicated by unknowns.

22

EXERCISE 2D

1 A book rests on a table. The magnitude of the normal contact force on the book from the table is 28 N. What is the mass of the book?

(M) 2 An oil drum of mass 250 kg rests on the ground. A vertical cable is attached to the drum and the tension is gradually increased. At one stage the tension in the cable has magnitude 1800 N. What is the magnitude of the normal contact force between the drum and the ground at this instant? What happens when the tension reaches 2500 N?

3 A fork-lift truck is raising a container of car batteries with an acceleration of $1.5\,\mathrm{m\,s^{-2}}$. The normal contact force on the container from the horizontal forks is 1610 N. Calculate the mass of the load.

4 A jet aircraft of mass 7 tonnes stands at rest on a part of the deck of an aircraft carrier that can be lowered to allow the jet to be housed in the hold of the carrier. Find the magnitude of the normal contact force on the wheels of the aircraft from the lowering part of the deck as it is lowered with an acceleration of $0.4\,\mathrm{m\,s^{-2}}$.

(M) 5 When a man stands on bathroom scales placed on the floor of a stationary lift, the reading is 90 kg. While the lift is moving upwards, he finds that the reading is 86 kg. Account for this change and describe the motion of the lift at this time.

6 A man of mass M kg and his son of mass m kg are standing in a lift. When the lift is accelerating upwards with magnitude $1\,\mathrm{m\,s^{-2}}$ the magnitude of the normal contact force exerted on the man by the lift floor is 880 N. When the lift is moving with constant speed the combined magnitude of the normal contact forces exerted on the man and the boy by the lift floor is 1000 N. Find the values of M and m.

7 A bucket of mass 0.5 kg contains a bag of sand of mass 4.5 kg. The bucket is pulled vertically upwards with constant acceleration using a rope. The bucket rises 24 metres in 5 seconds, starting from rest. Find the contact force that the bucket exerts on the bag of sand.

(PS) 8 A pile of three identical books, each of mass m kg, sit one on top of the other. The lowest book is raised with acceleration $1\,\mathrm{m\,s^{-2}}$. Find the contact force between each pair of adjacent books.

END-OF-CHAPTER REVIEW EXERCISE 2

(M) 1 A car moves on a straight horizontal road, under the action of a constant driving force of magnitude 1360 N. It accelerates from rest to the speed of $12.6\,\mathrm{m\,s^{-1}}$ in 8 seconds.

   a Assuming that any resistance forces can be ignored, find the mass of the car.

   b How would your answer change if a resistance force was included?

2 A box of mass 13 kg slides across a rough horizontal floor with an initial speed of $2.6\,\mathrm{m\,s^{-1}}$ and moves in a straight line. It comes to rest after it has travelled 3.7 m. Find the magnitude of the frictional force between the box and the floor.

3 A particle, of mass 0.4 kg, is projected horizontally with initial speed $10\,\mathrm{m\,s^{-1}}$. The particle is slowed by a constant resistance of 2.5 N.

A second particle, of mass 0.5 kg, falls vertically from rest.

a How long does it take until the first particle comes to rest?

b If the particles are released at the same time, after how many seconds have they travelled the same distance?

4 As a load moves downwards at a constant speed of $2\,\mathrm{m\,s^{-1}}$ the tension in the cable supporting it is 6000 N. Calculate the tension in the cable when the load is moving downwards with an acceleration of $2\,\mathrm{m\,s^{-2}}$.

5 A balloon of total mass 680 kg is descending with a constant acceleration of $0.4\,\mathrm{m\,s^{-2}}$. Find the upthrust acting on the balloon. When the balloon is moving at $1.5\,\mathrm{m\,s^{-1}}$, enough ballast is released for the balloon to fall with a deceleration of $0.2\,\mathrm{m\,s^{-2}}$. Calculate:

a how much ballast was released

b the time for which the balloon continues to fall before it begins to rise.

P **6** A stone of mass m is released from rest on the surface of a tank of water of depth d. During the motion, the water exerts a constant resisting force of magnitude R. The stone takes

t seconds to reach the bottom of the tank. Show that $R = m\left(g - \dfrac{2d}{t^2}\right)$.

M **7** A construction worker drops a screwdriver of mass 0.15 kg into a tank of water 1 metre deep. It enters the water with speed of $8\,\mathrm{m\,s^{-1}}$, and when it hits the base of the tank it is moving at $9\,\mathrm{m\,s^{-1}}$.

a What forces, apart from its weight, act on the screwdriver in the water?

b Assuming that, in the water, the screwdriver falls with constant acceleration $a\,\mathrm{m\,s^{-2}}$, calculate a. Hence find the total force opposing the motion of the screwdriver in the water.

PS **8** An acrobat of mass m slides down a vertical rope of height h. For the first three-quarters of her descent she grips the rope with her hands and legs so as to produce a frictional force equal to five-ninths of her weight. She then tightens her grip so that she comes to rest at the bottom of the rope. Sketch a (t, v) graph to illustrate her descent, and find the frictional force she must produce in the last quarter. If the rope is 60 metres high, calculate:

a her greatest speed

b the time she takes to descend.

24

Chapter 3
Forces in two dimensions

- Resolve forces in two dimensions.
- Find resultants of more than one force in two dimensions.
- Use $F = ma$ in two directions.
- Find directions of motion and accelerations.

 TIP

Unless otherwise stated, g should be taken as $10\,\text{m}\,\text{s}^{-2}$.

3.1 Resolving forces in horizontal and vertical directions in equilibrium problems

WORKED EXAMPLE 3.1

A ball of mass M kg is suspended on two light ropes. The first rope makes an angle of $30°$ with the upward vertical, the tension in this rope is $20\,\text{N}$. The second rope makes an angle of $45°$ with the upward vertical. The ball hangs in equilibrium. Show that $M = \sqrt{3} + 1$.

Answer

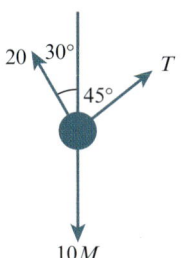

Draw a force diagram.

Tension in the second rope $= T\,\text{N}$.

Resolve the tension in each rope into horizontal and vertical components.

The ball is in equilibrium, so the horizontal resultant force $= 0$

$20 \sin 30° = T \sin 45°$

$\qquad T = 10\sqrt{2}$

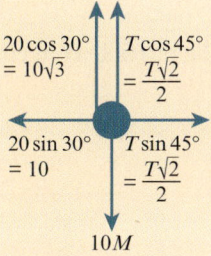

and the vertical resultant force $= 0$

$20 \cos 30° + T \cos 45° = 10\,M$

So, $10\sqrt{3} + 10 = 10\,M$

$\qquad M = \sqrt{3} + 1$

Substitute in the value found for the tension.

1 For each of the following diagrams, find the resolved parts of each of the forces in the directions Ox and Oy.

a

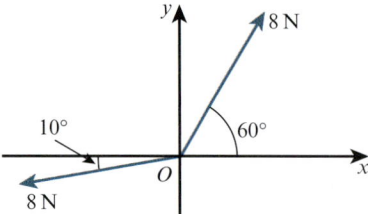

b

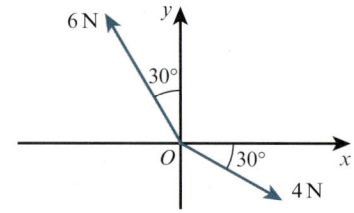

c

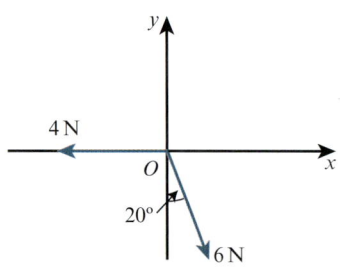

d

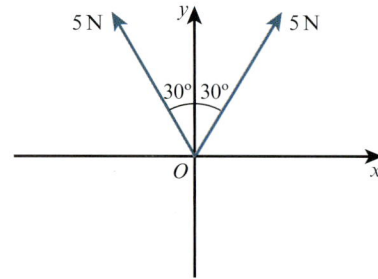

2 The diagram shows three coplanar forces acting on a particle P, which is in equilibrium. By resolving in the direction of the 5 N force, calculate the value of θ; by resolving in the direction of the X N force, calculate the value of X.

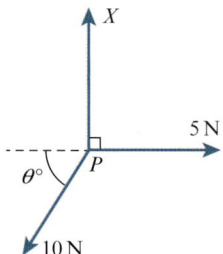

3 In the following diagram the particle is in equilibrium.

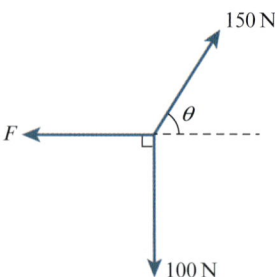

a Find the angle θ.
b Find the force F.

4   A mass of 200 g is being held by a string at an angle θ to the vertical with tension T newtons and a horizontal force of 5 N.

   a   Show that $T\cos\theta = 2$.

   b   Find the value of θ.

   c   Find the value of T.

5   A particle P of mass $4m$ kg is at rest on a horizontal table. A force of magnitude $50m$ N, acting upwards at an acute angle θ to the horizontal, is applied to the particle. Given that $\tan\theta = \dfrac{3}{4}$ and that there is a resistance to motion of magnitude $20m$ N, find the acceleration with which P moves. Find, in terms of m, the magnitude of the normal contact force of the table on P.

P 6   A lamp is supported in equilibrium by two chains fixed to two points A and B at the same level. The lengths of the chains are 0.3 m and 0.4 m and the distance between A and B is 0.5 m. Given that the tension in the longer chain is 36 N, show by resolving horizontally that the tension in the shorter chain is 48 N. By resolving vertically, find the mass of the lamp.

7   A block of wood of mass 5 kg rests on a table. A force of magnitude 35 N, acting upwards at an angle of θ to the horizontal, is applied to the block but does not move it. Given that the normal contact force between the block and the table has magnitude 30 N, calculate:

   a   the value of θ

   b   the frictional force acting on the block.

8   A man is pulling a chest of mass 40 kg along a horizontal floor with a force of 140 N inclined at 30° to the horizontal. His daughter is pushing with a force of 50 N directed downwards at 10° to the horizontal. The chest is moving with constant speed. Calculate the magnitude, F N, of the frictional force, and the magnitude, R N, of the normal contact force from the ground on the chest. Show that the ratio $\dfrac{F}{R}$ lies between 0.50 and 0.51.

PS 9   A particle is in equilibrium under the action of three forces acting in a horizontal plane, as shown in the diagram. The angle between the force F N and the force 4 N is 135°.

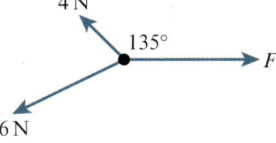

Find:

   a   the angle between the force 4 N and the force 6 N

   b   the magnitude of F.

10 A mass of 12 kg is suspended from two ropes. One rope makes an angle 30° with the upward vertical and the other makes an angle θ with the upward vertical. Calculate the tension in each rope, in terms of θ.

3.2 Resolving forces at other angles in equilibrium problems

WORKED EXAMPLE 3.2

A block of mass 5 kg sits on a slope inclined at 20° to the horizontal. The block is held in equilibrium and is prevented from sliding down the slope by a frictional force. The block is then pulled up the slope at a constant speed, using a rope that makes an angle of 50° to the horizontal. Assume that the frictional force has the same magnitude as before but now acts down the slope. Find the tension in the rope.

TIP

Resolve forces parallel and perpendicular to the slope.

Answer

Block about to slip down the slope:

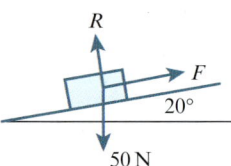

The component of the weight in the direction that is parallel to the slope is
$50 \sin 20° = 17.1\,\text{N}$ down the slope

Because the normal reaction and the frictional force are perpendicular, it is easier to resolve the weight in the directions parallel and perpendicular to the slope, rather than resolving R and F horizontally and vertically.

The angle between the slope and the horizontal is the same as the angle between the normal and the vertical.

The angle between the weight and the (inward) normal is 20°.

Resolving forces parallel to the slope

$F = 17.1\,\text{N}$

The force up the slope balances the component of the weight down the slope.

Block being pulled up the slope.

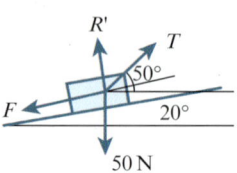

The angle between the slope and the horizontal is 20°, the angle between the rope and the horizontal is 50°, so the angle between the rope and the slope is 30°.

Resolving forces parallel to the slope

$F + 50 \sin 20° = T \cos 30°$
$T = (17.1 + 17.1) \div \cos 30°$
$\quad = 39.5\,\text{N}$

$F = 17.1$ from the previous part.

Component of the weight down the slope

$= 50 \sin 20° = 17.1\,\text{N}$ as before.

EXERCISE 3B

1   Calculate the magnitude of the horizontal force needed to maintain a crate of mass 6 kg in equilibrium, when it is resting on a frictionless plane inclined at 20° to the horizontal. Calculate also the magnitude of the normal contact force acting on the crate.

2   A skier of mass 78 kg is pulled at constant speed up a slope, of inclination 12°, by a force of magnitude 210 N acting upwards at an angle of 20° to the slope (see diagram). Find the magnitudes of the frictional force and the normal contact force acting on the skier.

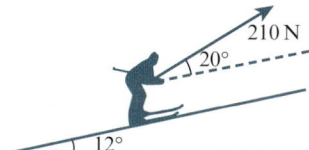

M   3   A particle of weight 10 N is placed on a smooth plane inclined at 35° to the horizontal. Find the magnitude of the force required to keep the particle in equilibrium if it acts:

a   parallel to the plane

b   horizontally

c   upwards at an angle of 25° to a line of greatest slope of the plane.

Without making any calculations state, with a reason, which of the three cases has the greatest normal contact force.

4   A metal sphere of weight 500 N is suspended from a fixed point O by a chain. The sphere is pulled to one side by a horizontal force of magnitude 250 N and the sphere is held in equilibrium with the chain inclined at an angle θ to the vertical. Find, in either order, the tension in the string and the value of θ.

5   Three friends are pulling a 100 kg load across a smooth horizontal surface. Omar pulls with a force of 200 N. Ren pulls with a force of 150 N and is positioned at an angle 15° clockwise from Omar. David pulls with force k N and is positioned at an angle 40° clockwise from Ren.

Modelling the load as a particle, and given that the load begins to move exactly towards Ren, find k and determine the initial acceleration of the load.

M   6   A small, smooth ring R of weight 5 N is threaded on a light, inextensible, taut string. The ends of the string are attached to fixed points A and B at the same horizontal level. A horizontal force of magnitude 4 N is applied to R. In the equilibrium position the angle ARB is a right-angle, and the portion of the string attached to B makes an angle $\theta°$ with the horizontal.

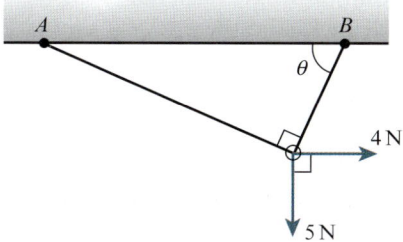

a   Explain why the tension T is the same in each part of the string.

b   Find T and θ.

7   A boat on a trailer is held in equilibrium on a slipway inclined at 20° to the horizontal, by a cable inclined at $\theta°$ to a line of greatest slope of the slipway, as shown in the diagram. The combined mass of the boat and trailer is 1250 kg. The cable may break if the tension exceeds 7000 N. Find the maximum value of θ.

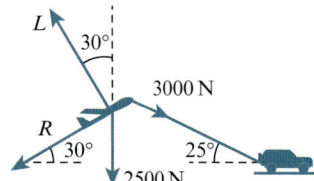

PS   8   A glider of weight 2500 N is being towed with constant speed by a four-wheel drive vehicle. The towrope is inclined at 25° to the horizontal and the glider is inclined at 30° to the horizontal as shown in the diagram. The air resistance has magnitude R N and the lift has magnitude L N; the directions in which they act are shown in the diagram. Calculate the values of R and L, given that the tension in the rope is 3000 N.

E   **3.3 The triangle of forces and Lami's theorem for three-force equilibrium problems**

> **TIP**
>
> The methods in this section are not required by the syllabus, but can be useful when solving problems.

WORKED EXAMPLE 3.3

A particle is held in equilibrium under the action of three forces acting in a horizontal plane, as shown in the diagram.

The angle between the F_1 N force and the 2 N force is 95° and the angle between the 2 N force and the F_2 N force is 105°. Find the magnitude of F_1 and the magnitude of F_2.

Answer

Draw a triangle of forces.

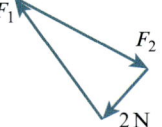

The forces must join 'nose to tail' around the triangle, but can be drawn in any order.

Using the sine rule:

$$\frac{2}{\sin 20°} = \frac{F_1}{\sin 75°} = \frac{F_2}{\sin 85°}$$

$F_1 = 5.65, \; F_2 = 5.83$

The internal angle between the 2 N force and force F_1 is $180° - 95° = 85°$.

Similarly the internal angle between the force F_2 and the 2 N force is 75°.

The internal angle between the force F_1 and the force F_2 is 20°.

Or using Lami's theorem:

$$\frac{2}{\sin 160°} = \frac{F_1}{\sin 105°} = \frac{F_2}{\sin 95°}$$

$F_1 = 5.65, \; F_2 = 5.83$

The angle 'opposite' F_2 is 95°. The angle opposite F_1 is 105° and the angle opposite the 2 N force is $360° - 200° = 160°$.

EXERCISE 3C

1 A box sits in equilibrium on a rough inclined plane. The plane makes an angle $\theta°$ with the horizontal. The frictional force is $0.3 \times$ the normal contact force. Find the value of θ.

2 A particle is held in equilibrium under the action of three forces acting in a horizontal plane. The angle between the first force and the second force is 120°. The angle between the second force and the third force is 80° and the magnitude of the third force is 17 N. Find the magnitude of the first force.

3 A sign is suspended from two wires. The first wire makes an angle of 50° to the horizontal and the tensions in the wires are 3 N and 4 N, as shown in the diagram. Find the mass of the sign.

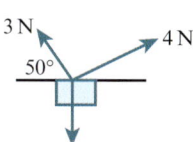

4 An object is in equilibrium under the action of three horizontal forces **P**, **Q** and **R**. Use a triangle of forces to find the magnitudes of **Q** and **R** respectively in each of the following cases.

   a **P** has magnitude 10 N and bearing 090°, **Q** and **R** have bearings 210° and 340° respectively

   b **P** has magnitude 20 N and bearing 020°, **Q** and **R** have bearings 090° and 240° respectively
   Confirm your answers by resolving in directions perpendicular to **R** and **Q**.

5 A cable car is connected to a cable by two rigid supports which make angles of 75° and 35° with the upward vertical, as shown in the diagram. Find the tensions in the supports when the total weight of the cable car and its passengers is 8000 N and the cable car is stationary.

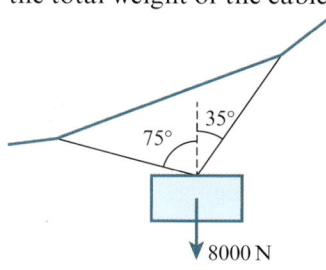

P 6 A boat is held in equilibrium by three horizontal ropes. Each rope is in tension. The angle between the first rope and the second rope is twice the angle between the second rope and the third rope.

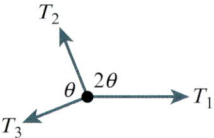

Use the identity $\sin 2\theta = 2\sin\theta\cos\theta$ to deduce that if tension in the first rope equals the tension in the third rope then the angle between the first rope and the second rope is $120°$ and explain why this is not physically possible.

3.4 Non-equilibrium problems for objects on slopes and known directions of acceleration

WORKED EXAMPLE 3.4

A raft of mass 50 kg is pulled along a canal using two ropes. The rope to the left bank of the canal has a tension of 100 N and is at an angle of $20°$ to the direction of motion. The rope to the right bank of the canal is at an angle of $25°$ to the direction of motion. The resistance to the motion of the raft is 40 N. Find the acceleration of the raft.

Answer

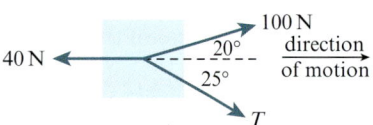

Resolving perpendicular to the direction of motion

$100\sin 20° = T\sin 25°$

so $T = 80.93\,\text{N}$

The resisting force has no component in this direction, so the components of the two tensions must therefore be equal.	
Use more than 3 significant figures in working.	

Resultant force along the direction of motion

$= 100\cos 20° + T\cos 25° - 40$

$= 127.3\,\text{N}$

The resisting force is negative.

Substitute in the value of the tension found to find the resultant force.

$F = ma$

Use Newton's second law.

$127.3 = 50a$

$a = 2.55\,\text{m s}^{-2}$

Solve to find the acceleration and now round to 3 significant figures.

32

EXERCISE 3D

1   A toy of mass 1.8 kg is pulled along a horizontal surface by a string inclined at 30° to the horizontal. Given that the tension in the string is 6 N and that there are no forces resisting motion, calculate the acceleration of the toy.

2   A van of mass 750 kg is being towed along a horizontal road with constant acceleration 0.8 m s^{-2} by a breakdown vehicle. The connecting towbar is inclined at 40° to the horizontal. Given that the tension in the towbar has magnitude 1000 N, calculate the magnitude of the force resisting the motion of the van.

3   A shopper pushes a shopping trolley in a straight line towards her car with a force of magnitude 20 N, directed downwards at an angle of 15° to the horizontal. Given that the acceleration of the trolley is 2.4 m s^{-2}, calculate its mass. Find also the magnitude of the normal contact force exerted on the trolley by the ground.

P   4   The diagram shows three horizontal forces acting on a mass of 4 kg. Given that the mass moves in the direction of the dotted line, show that $\theta = 30°$ and find the magnitude of the acceleration.

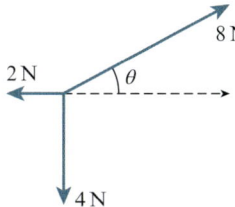

5   A paraglider of mass 90 kg is pulled by a rope attached to a speedboat. With the rope making an angle of 20° to the horizontal, the paraglider is moving in a straight line parallel to the surface of the water with an acceleration of 1.2 m s^{-2}. The tension in the rope is 250 N. Calculate the magnitude of the vertical lift force acting on the paraglider, and the magnitude of the air resistance.

6   The lid of a desk is hinged along one edge so that it can be tilted at various angles to the horizontal. A book of mass 1.8 kg is place on the lid. The lid is tilted to an angle of 15°, as shown in the diagram. Find the frictional force, given that the book does not move.

As the desk lid is tilted further, the book begins to move when the lid is inclined at $\theta°$ to the horizontal. Given that the maximum magnitude that the frictional force can attain is 8.45 N, find the maximum value of θ for which equilibrium remains unbroken.

The book is held at rest and the lid is tilted until it is inclined at 40° to the horizontal. The book is then released. Find the acceleration of the book down the desk lid, assuming the maximum frictional force still acts.

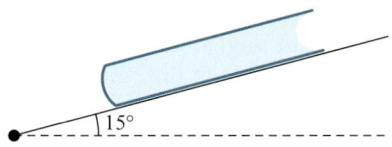

33

7   A box of mass 12 kg is dragged, with a constant acceleration of $1.75\,\mathrm{ms^{-2}}$, up a path inclined at $30°$ to the horizontal. The force pulling the box has magnitude $2X\,\mathrm{N}$ and acts at $10°$ to the path, as shown in the diagram. The frictional force has magnitude $X\,\mathrm{N}$. Calculate the value of X and the magnitude of the normal contact force of the path on the box.

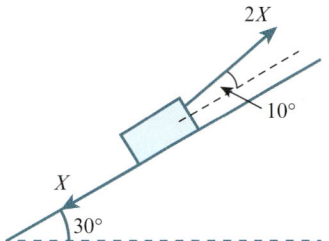

3.5 Non-equilibrium problems and finding resultant forces and directions of acceleration

WORKED EXAMPLE 3.5

Two students are moving a crate of mass 400 kg across a rough floor. One student pulls the crate with a force of 40 N and the other student pushes the crate with a force of 50 N. The angle between these forces is $120°$, as shown in the diagram.

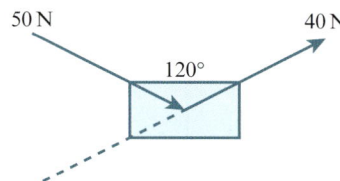

a   Calculate the angle between the direction in which the first student pulls and the direction in which the crate moves.

b   Calculate how long it takes to move the crate 5 metres, starting from rest.

Answer

a

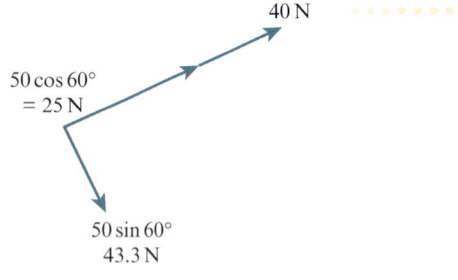

Resolve the 50 N force into a component along the direction of the 40 N force and a component perpendicular to this direction.

$180° - 120° = 60°$

The resultant force has a component of 65 N along the direction of the 40 N force and a component of 43.3 N perpendicular to this direction.

θ = angle between the direction in which the first student pulls and the direction in which the crate moves.

$\tan\theta = \dfrac{43.3}{65} = 0.666$

$\theta = 33.7°$

Consider the right-angled triangle below, showing the direction of motion of the crate.

b Magnitude of the resultant force

$= \sqrt{65^2 + 43.3^2}$ .... Use Pythagoras in the same right-angled triangle.

$= 78.1\,\text{N}$ .... Find the resultant force in order to find the acceleration of the crate.

$F = ma$ .... Use Newton's second law.

$78.1 = 400\,a$

$a = 0.195\,\text{m s}^{-2}$

$s = 5,\ u = 0,\ a = 0.195$ .... List the known values.

$s = ut + \dfrac{1}{2}at^2$ .... Choose the most suitable equation of constant acceleration.

$5 = \dfrac{1}{2} \times 0.195\,t^2$ .... Substitute in the known values.

$t = 7.16\,\text{seconds}$ .... Rearrange to find the time taken.

EXERCISE 3E

1 A small boat is moving under the action of four forces: a force of magnitude 50 N on a bearing of 030°, a force of magnitude 90 N on a bearing of 100°, a force of magnitude 70 N on a bearing of 200° and a force of magnitude 30 N on a bearing of 330°.

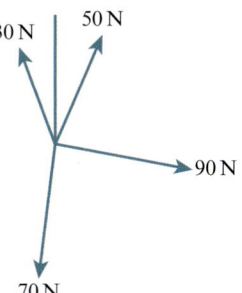

Find the bearing, to the nearest degree, of the direction in which the boat moves.

2 Three coplanar forces act on a particle of mass 0.5 kg. The first force has magnitude 5 N and acts at 45° to the positive x-axis and at 45° to the positive y-axis. The second force has magnitude 6 N and acts along the negative x-axis. The third force has magnitude 4 N and acts at an angle of 45° to the positive x-axis and at 45° to the negative y-axis. Calculate the magnitude of the acceleration of the particle.

3 The resultant vertical force acting on a hot air balloon is 50 N vertically upwards. A wind of magnitude 30 N acts on the balloon at an angle of 60° to the vertical. Find the possible angles between the resulting direction of motion of the balloon and the upward vertical.
Hint: consider the possible directions of the wind.

4 The drive force from the engine of a light aircraft is 400 N. The pilot sets a course for the plane to fly due North. A strong wind of 40 N blows from West to East and pushes the plane off its course. What bearing, to the nearest degree, should the pilot set for the plane to actually fly due North?

5 A floating buoy is moved using three wires. Each of the wires acts horizontally. The angle between the first and second wires is 40°, the angle between the second and third wires is 200° and the angle between the third and first wires is 120°. The tension in the first wire is twice the tension in the second wire and three times the tension in the third wire. Find, to 1 decimal place, the angle between the direction of motion and the first wire.

END-OF-CHAPTER REVIEW EXERCISE 3

1 A particle P is in equilibrium on a smooth horizontal table under the action of horizontal forces of magnitudes F N, $2F$ N, G N and 8 N, acting in the directions shown. Find the values of F and G.

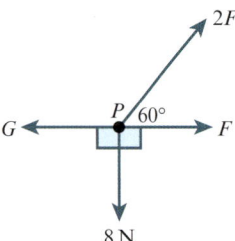

2 A small block is held at rest on a smooth plane inclined at 30° to the horizontal. The particle is held in equilibrium by a horizontal force of magnitude 20 N acting in the vertical plane containing the line of greatest slope.

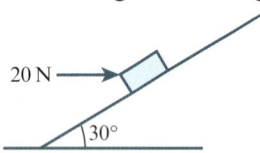

Find the weight of the block.

3 Three horizontal forces of magnitudes 10 N, F N and $2F$ N act at a fixed point and are in equilibrium.

The angle between the 10 N force and the F N force is $\theta°$, the angle between the F N force and the $2F$ N force is 90° and the angle between the $2F$ N force and the 10 N force is $270° - \theta°$.

Calculate the angle $\theta°$, to the nearest degree, and the value of F.

4 A cyclist exerts a constant driving force of magnitude 80 N while cycling along a straight horizontal road against a constant resistance of F N. The total weight of the cyclist and her bicycle is 700 N. The cyclist's acceleration is $0.1 \, \text{m s}^{-2}$.

a Find the value of F.

The cyclist then travels with the same driving force and against the same resistance as before, up a hill inclined at an angle α to the horizontal.

The cyclist's acceleration is $0.01 \, \text{m s}^{-2}$ in the direction of travel, up the hill.

b Find the value of $\sin \alpha$.

5 A particle P of mass 0.2 kg lies on a smooth horizontal plane. Horizontal forces of magnitudes 0.3 N, 0.4 N, 1.2 N and 1.3 N act on P. The 0.3 N force is perpendicular to the 0.4 N force and P is in equilibrium.

The 1.3 N force is then removed.

   a Find the magnitude of the acceleration with which P begins to move.

   b Find the possible angles between the direction of motion and the 0.3 N force, giving your answers as angles between $0°$ and $180°$.

M   6 A box of mass 4 kg is held at rest on a plane inclined at an angle of $2.5°$ to the horizontal. The box is then released and slides down the plane.

   a A simple model assumes that the only forces acting on the box are its weight and the normal reaction from the plane. Show that, according to this simple model, the acceleration of the box down the plane is $0.436 \, \text{m s}^{-2}$, correct to three significant figures.

   b In fact, the box moves down the plane with constant acceleration and takes 2 s to travel 0.8 m. By using this information, find the acceleration of the box.

   c Explain why the answer to part **b** is less than the answer to part **a**.

7 Three horizontal forces of magnitudes 15 N, 11 N and 13 N act on a particle P in the directions shown in the diagram.

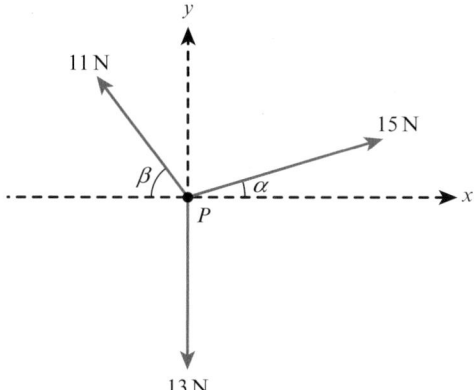

The particle is in equilibrium.

Find the value of $\sin \alpha$.

8 A block with mass 2 kg is pulled from rest up a smooth slope inclined at $20°$ to the horizontal by a string with tension T, maintained at an angle of $30°$ to the horizontal.

   a After 5 seconds, the block has moved 1 m along the slope. Calculate T.

   b The block is allowed to come to rest again, then the tension is increased so that the block is about to lift off the slope. Calculate the minimum tension needed to achieve this.

Chapter 4
Friction

- Calculate the size of frictional forces.
- Use friction to solve problems in motion.
- Determine the direction of motion of an object.
- Solve problems where a change in direction of motion changes the direction of friction.

4.1 Friction as part of the contact force

> 💡 **TIP**
>
> Unless otherwise stated, g should be taken as $10\,\text{m}\,\text{s}^{-2}$.

WORKED EXAMPLE 4.1

A car of mass $1500\,\text{kg}$ has broken down on a slope that makes an angle of $8°$ to the horizontal. The brakes of the car are off and the driver is trying to push the car down the slope. The driver pushes with a force of $400\,\text{N}$ parallel to the slope but because of friction the car does not move.

    a   Modelling the car as a particle, find the angle that the total contact force makes with the slope.

    b   Find the minimum value of the coefficient of friction between the car and the slope.

Answer

a

push 400 N · R normal contact force · F frictional resistance · 8° · weight 15 000 N

Draw a force diagram.

Weight = $10 \times 1500 = 15\,000\,\text{N}$.

Resolve the weight into components parallel and perpendicular to the slope.

Angle between the weight and the normal to the slope = $8°$.

$15\,000 \sin 8°$
$= 2090\,\text{N}$

$15\,000 \cos 8° = 14\,850\,\text{N}$

Resolving perpendicular to the slope
$R = 14\,850\,\text{N}$
Resolving parallel to the slope
$F = 400 + 2090 = 2490\,\text{N}$

Car does not move so forces are in equilibrium.

θ = angle between the total contact force and the slope

$$\tan \theta = \frac{R}{F}$$
$$\theta = 82°$$

b Minimum value for the coefficient of friction
$$F = \mu R$$
$$\mu = \frac{F}{R} = 0.141$$

Total contact force is the resultant of R and F. Use exact values in the calculation.

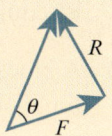

Note this is also $90° - 8°$.

Use the exact values found; the minimum value for the coefficient of friction is if the car is just about to move at this point.

EXERCISE 4A

1 The diagram shows horizontal forces of magnitudes P N and Q N acting in opposite directions on a block of mass 5 kg, which is at rest on a horizontal surface. State, in terms of P and Q, the magnitude and direction of the frictional force acting on the block when:

a $P > Q$ b $Q > P$.

2 The diagram shows horizontal forces of magnitudes P N and 100 N acting in opposite directions on a block of weight 50 N, which is at rest on a horizontal surface. Given that the coefficient of friction between the block and the surface is 0.4, find the range of possible values of P.

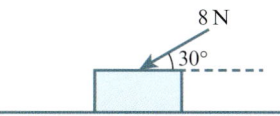

3 The diagram shows a force of magnitude 8 N acting downwards at 30° to the horizontal on a block of mass 3 kg, which is at rest on a horizontal surface. Calculate the frictional force on the block.

4 In each of the following problems, a block of mass m kg is pulled along a rough horizontal surface by a light horizontal rope with tension T newtons. The acceleration is a m s^{-2}. The coefficient of friction is μ.

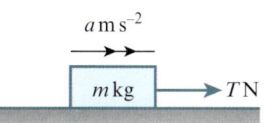

a Find μ when:

  i $m = 2$, $T = 5$, $a = 0.1$ ii $m = 8$, $T = 2$, $a = 0.2$

b Find a if:

  i $m = 1.1$, $T = 6$, $\mu = 0.455$ ii $m = 5$, $T = 1$, $\mu = 0.01$

39

5 A car of mass 1000 kg has a coefficient of friction of 0.9 with the road. What horizontal force is required to move the car?

6 A block of mass 1.5 kg lies in limiting equilibrium on a horizontal surface, with a horizontal force of 6 N applied to it.

    a Find the coefficient of friction between the block and the surface.

    b Find the magnitude of the contact force between the block and the surface.

4.2 Limit of friction

> **TIP**
>
> If the object is moving relative to the surface, friction will take the value $F = \mu R$.

WORKED EXAMPLE 4.2

A box of mass 200 kg is dragged across a rough horizontal surface by a chain attached to a truck. The angle between the chain and the horizontal is 30°. The coefficient of friction between the box and the ground is 0.32. Calculate the tension needed in the chain to enable the box to move.

Answer

normal contact force R

tension, T

$\mu = 0.32$

friction F

30°

weight = 2000 N

Model the box as a particle and the chain as light and straight.

It is important to be clear that F is the frictional force and not the resultant force (as in $F = ma$).

Resolve vertically:

$R + T\sin 30° = 2000$

so $R = 2000 - 0.5\,T$

Resolve vertically to find R in terms of T.

Resultant horizontal force

$= T\cos 30° - F$

The box moves if there is a resultant horizontal force in the direction of motion.

$F = \mu R$

$= 0.32\,R$

The box is moving so friction is limiting.

So $F = 0.32(2000 - 0.5T) = 640 - 0.16\,T$

Substitute $R = 2000 - 0.5\,T$ in $F = 0.32\,R$.

Resultant horizontal force

$= 0.866\,T - (640 - 0.16\,T)$

$= 1.026\,T - 640$

Substitute $F = 640 - 0.16\,T$ in $T\cos 30° - F$.

$1.026\,T - 640 > 0$

$T > 624\,\text{N}$

So the tension must be greater than 624 N for the box to move.

For motion, the resultant horizontal force must be positive.

40

EXERCISE 4B

1   An airline passenger pushes a 15 kg suitcase along the floor with his foot. A force of 60 N is needed to move the suitcase. Find the coefficient of friction. What force would be needed to give the suitcase an acceleration of $0.2\,\mathrm{m\,s^{-2}}$?

2   A block of mass 6 kg is accelerating at $1.25\,\mathrm{m\,s^{-2}}$, on a horizontal surface, under the action of a horizontal force of magnitude 22.5 N. Calculate the coefficient of friction between the block and the surface.

 3   The diagram shows a block of mass 4 kg at rest on a plane inclined at 35° to the horizontal, under the action of a force of magnitude P N acting up the plane. The coefficient of friction between the block and the plane is 0.45. Find the range of possible values of P.

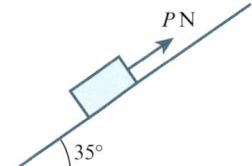

4   A child pulls a toybox of mass 3.5 kg across a rough floor, using a light string tied to the box at one end. The tension in the string is 28 N and the string remains at an angle of 25° to the horizontal. If the coefficient of friction between the box and the floor is 0.7, what is the acceleration of the box?

> **TIP**
>
> To calculate the acceleration, resolve perpendicular to the surface first to find the normal contact force and, hence, the frictional force. Then resolve parallel to the surface and calculate acceleration using Newton's second law.

5   A car of weight 12 000 N, travelling at $25\,\mathrm{m\,s^{-1}}$, skids to a halt in 50 m, taking 4 seconds.

   a   Assuming a constant braking force, find the deceleration of the car.

   b   Assume that friction is the only force acting on the car. Find the coefficient of friction between the car and the road.

6   A particle of mass 5 kg rests on a rough horizontal table. The coefficient of friction between the particle and the table is 0.4. A light inextensible string, inclined at an angle 20° above the horizontal, is attached to the particle. The tension in the string is 15 N.

   a   Show that the particle remains at rest.

   b   Find the magnitude of the contact force between the particle and the table.

 7   A particle of mass m lies on a rough horizontal surface. The coefficient of friction between the particle and the surface is μ. A horizontal force acts on the particle, which is in limiting equilibrium. Show that the magnitude of the contact force between the particle and the surface is $mg\sqrt{1 + \mu^2}$.

P   8   A particle of mass 1 kg lies on a rough horizontal surface, with the
coefficient of friction between surface and particle equal to 0.75. A light
inextensible string is attached to the particle, and tension applied with
the string at an angle θ above the horizontal where $0° < \theta < 90°$.

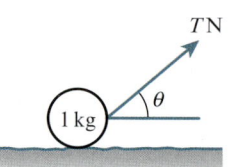

Given the system is in limiting equilibrium, show that the tension T
in the string satisfies the equation $T = \dfrac{5.88}{\sin(\theta + \alpha)}$ for some value α, and find α.

4.3 Change in direction of friction in different stages of motion

WORKED EXAMPLE 4.3

A crate of weight 500 N is being winched up a slope using a rope.
The slope makes an angle of 10° to the horizontal and the rope
is parallel to the slope. The tension in the rope is 200 N and the
coefficient of friction between the crate and the slope is 0.05.
The crate starts from rest and has been pulled 5 metres along the
slope when the rope breaks. The crate continues to move up the
slope and then slides back down the slope.

How long does it take from when the crate starts to be winched up
the slope to when it returns to the start point?

Answer

Travelling up the slope before
the rope breaks:

normal contact force R

$\mu = 0.05$ →   tension 200 N

friction F
                10°

weight 500 N

Resolving perpendicular to the slope:

$R = 500 \cos 10° = 492$ N

$F = \mu R$
$\quad = 0.05 \times 492 = 24.6$ N

Resultant parallel to the slope
(up the slope)
$\quad = 200 - 500 \sin 10° - F$
$\quad = 88.6$ N

TIP

Draw a different
force diagram for
each stage of the
motion and deal
with the stages
separately. The
direction of the
frictional force
will be different if
the object changes
direction.

It is important to be clear
that F is the frictional force
and not the resultant force
(as in $F = ma$).

Resolve perpendicular to
the slope to find the normal
contact force.

The crate is moving so
friction is limiting.

Resolve in the direction of
motion to find the resultant
force.

Resultant force = mass × acceleration

> Weight = 10 × mass.

$88.6 = 50a$

> Use Newton's second law.

$a = 1.77\,\mathrm{m\,s^{-2}}$

> Constant acceleration up the slope.

When the rope breaks:
$s = 5,\ u = 0,\ a = 1.77$

> State the values for the first section of motion.

$v^2 = u^2 + 2as$ so $v = 4.21\,\mathrm{m\,s^{-1}}$

> Use the most suitable equation of motion to find the velocity when the rope breaks.

$v = u + at$ so $t = 2.38\,\mathrm{s}$

Rope breaks after 2.38 seconds, the crate is then moving at $4.21\,\mathrm{m\,s^{-1}}$ up the slope.

> Use the most suitable equation of motion to find the time taken for the rope to break.

Travelling up the slope after the rope breaks:

normal contact force R

friction F

weight 500 N

> Draw a new diagram of forces, this time omitting the tension in the rope.

Resolving perpendicular to slope
$R = 500 \cos 10° = 492\,\mathrm{N}$

> Resolve perpendicular to the slope to find the normal contact force, note this is the same as before.

$F = \mu R = 24.6\,\mathrm{N}$

> Again the crate is moving so friction is limiting.

Resultant force parallel to the slope (up the slope)
$= -500 \sin 10° - F$
$= -111\,\mathrm{N}$

> Resolve in the direction of motion to find the resultant force, note this has changed now there is no tension in the rope.

Resultant force = mass × acceleration
$-111 = 50a$
$a = -2.22\,\mathrm{m\,s^{-2}}$

> Use Newton's second law.

> The crate is now decelerating up the slope.

From when the rope breaks until when the crate comes to instantaneous rest:

$u = 4.21$, $v = 0$, $a = -2.22$ .... State the values for the second section of motion.

$v^2 = u^2 + 2as$ so $s = 3.97\,\text{m}$ .... Use the most suitable equation of motion to find the distance the crate travels up the slope during the second stage of motion.

$v = u + at$ so $t = 1.89\,\text{s}$ .... Use the most suitable equation of motion to find the time taken for the second stage of motion.

$2.38 + 1.89 = 4.26\,\text{s}$ .... Add the two times for the two sections of motion.

$5 + 3.97 = 8.97\,\text{m}$ .... Add the two distances for the two sections of motion.

The crate comes to instantaneous rest 4.26 seconds after starting, crate is then 8.98 m up the slope from the start.

Sliding down the slope: .... Friction opposes motion so it now acts up the slope.

normal contact force R

friction F

weight 500 N

Although the same symbol has been used, the frictional force is not the same as before.

Resolving perpendicular to slope: .... Resolve perpendicular to the slope to find the normal contact force, note this is the same as before.

$R = 500 \cos 10° = 492\,\text{N}$

$F = \mu R = 24.6\,\text{N}$ .... Again the crate is moving so friction is limiting.

Resultant force parallel to the slope (down the slope)
$= 500 \sin 10° - F$ .... Resolve in the direction of motion to find the resultant force, note the weight component is now positive as the crate is moving down the slope.
$= 62.2\,\text{N}$

Resultant force = mass × acceleration ... Use Newton's second law.

$62.2 = 50\,a$

$a = 1.24\,\mathrm{m\,s^{-2}}$ · · · · · · · · · · · The crate is now accelerating down the slope.

From when the crate comes to instantaneous rest to when it returns to the start point:

$s = 8.97,\ u = 0,\ a = 1.24$ · · · · · · · · State the values for the final section of motion.

$s = ut + \dfrac{1}{2}at^2$

$t = 3.80\,\mathrm{s}$ · · · · · · · · · · · · · · Use the most suitable equation of motion to find the time taken for the final stage of motion.

$3.80 + 4.26 = 8.06\,\mathrm{s}$ · · · · · · · · Add this time to the previous time found for the first and second sections of motion.

The crate returns to the start 8.06 seconds after starting.

EXERCISE 4C

1 A particle P is projected upwards along a line of greatest slope from the foot of a surface inclined at 45° to the horizontal. The initial speed of P is $8\,\mathrm{m\,s^{-1}}$ and the coefficient of friction is 0.3. The particle P comes to instantaneous rest before it reaches the top of the inclined surface.

    a Calculate the distance P moves before coming to rest.

    b Calculate the time P takes before coming to rest.

    c Find the time taken for P to return to its initial position from its highest point.

PS 2 A block B of mass 1 kg lies on a smooth plane inclined at 35° to the horizontal. A light, inextensible string is attached at one end to B, and runs from B upslope parallel to the line of greatest slope on the plane to a smooth peg. This string is attached at the other end to a particle P of mass 2 kg, which hangs vertically below the peg, exactly 1 metre from the floor. The system is released from rest.

    a Find the acceleration of B up the slope.

    b When P hits the floor, the string breaks. Assuming the initial distance between the block and the peg is sufficiently great that the block will not reach the peg, find the total distance travelled by the block when it instantaneously passes through its initial position.

3 An ice-hockey puck is struck from one end of a rink of length 27 m towards the other end. The initial speed is $6\,\mathrm{m\,s^{-1}}$, and the puck rebounds from the boundary fence at the other end with a speed which is 0.75 times the speed with which it struck the fence, before just returning to its starting point. Calculate the coefficient of friction between the puck and the ice.

4   A particle P with mass 4.5 kg lies on a rough plane inclined
    at 30° to the horizontal. A light, inextensible string connects
    to P then runs parallel with the line of greatest slope of the
    plane to a smooth peg, then vertically downwards, through
    a smooth, free ring R with mass 2 kg and then vertically
    upwards to attach to a fixed point S.

    The coefficient of friction between P and the plane is 0.15.

    a   By resolving forces vertically at R, show that the
        acceleration of R when the system is released from
        rest is related to the tension T in the string by $a + T = g$.

    b   By resolving forces at P, find an equation linking a and T with friction F.

    c   Find the direction and magnitude of the frictional force.

    d   Determine whether P will remain stationary, move upslope or move downslope when the
        system is released from rest.

5   A toy hovercraft with mass 0.5 kg is placed on a rough surface
    inclined at 30° to the horizontal. Propulsion of the toy is achieved
    by directing two small fans, one either side of the toy, set at an
    angle to provide a driving force for movement. The force produced
    by the fans is 10 N.

    The coefficient of friction between the toy and the surface is 0.1.

    a   Find the acceleration of the toy up the slope if the fans are set
        to blow horizontally.

    b   Find the acceleration of the toy up the slope if the fans are set to blow parallel to the slope.

    c   The fans are set so that the direction of the driving force is an angle θ greater than the
        slope (i.e. at $(30° + \theta)$ above the horizontal). Find the value of θ that maximises the
        acceleration.

6   The diagram shows a force of magnitude 20 N acting downwards
    at 25° to the horizontal on a block of mass 4 kg, which is at rest in
    limiting equilibrium on a horizontal surface. Calculate the
    coefficient of friction between the block and the surface.

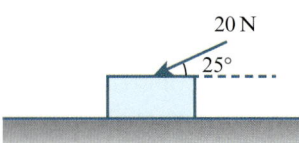

    The direction of the force of magnitude 20 N is now reversed.
    Calculate the acceleration with which the block starts to move.

7   A particle is projected upwards on a rough slope inclined at an angle θ to the horizontal.
    There is a coefficient of friction μ between the particle and the slope. The acceleration on the
    way up is twice the acceleration on the way down. Prove that $\tan \theta = 3\mu$.

4.4 Angle of friction

> **TIP**
>
> The methods in this section are not required by the syllabus, but can be useful when solving problems.

WORKED EXAMPLE 4.4

A book of mass 0.75 kg sits in the middle of a horizontal plank. The plank is 1.3 metres long. One end of the plank is gradually raised, with the other end remaining fixed, so the angle between the plank and the horizontal is gradually increased. When the book starts to slip the end of the plank has been raised through a height of 50 cm. Calculate the coefficient of friction between the plank and the book.

> **TIP**
>
> The angle of friction is related to the coefficient of friction by $\lambda = \tan^{-1}\mu$.

Answer

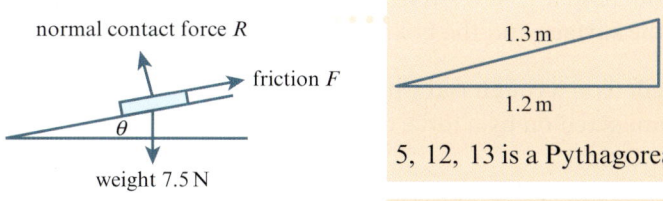

5, 12, 13 is a Pythagorean triple.

In limiting equilibrium: θ = angle of friction.

$$\mu = \tan\theta = \frac{5}{12}$$

EXERCISE 4D

  1 A bowl of mass 500 grams is placed on a table, which is tilted at various angles to the horizontal. The coefficient of friction is 0.74. Calculate the net force on the bowl down a line of greatest slope on the table when the angle of tilt is:

    a $36°$ **b** $36\frac{1}{2}°$ **c** $37°$.

    Describe what happens in each of these cases.

  2 A block of mass 3 kg lies at rest on a rough board, with $\mu = 0.45$. If the board is slowly raised at one end, beyond what angle will the block begin to slide?

  3 A block of mass m lies on a flat, rough surface. The coefficient of friction between the block and the surface is μ. The surface is initially horizontal, and its inclination is gradually increased. When the inclination of the surface exceeds $42°$ to the horizontal, the block begins to move.

    a Find μ.

    b The inclination of the surface is increased further to $44°$. Find, in terms of m, the magnitude of the contact force between the block and the surface.

47

4 A block B of mass $3\,\text{kg}$ lies on a surface inclined at $45°$ to the horizontal. A light, inextensible string is attached at one end to B, and runs from B upslope parallel to the line of greatest slope on the plane to a smooth peg P. The string passes over the peg, through a smooth ring R of mass $4\,\text{kg}$, and is attached to a wall at W.

Given that the angle PRW equals $120°$ and the system is in equilibrium, determine the possible values for μ, the coefficient of friction between block B and the surface.

END-OF-CHAPTER REVIEW EXERCISE 4

1 A box of weight $300\,\text{N}$ sits on a rough horizontal surface. A string is attached to the box and a tension of magnitude $X\,\text{N}$ is applied to the box at an angle of θ to the vertical. The box remains in equilibrium.

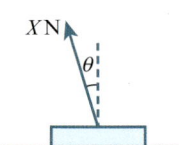

a Find, in terms of X and θ, the normal component of the force exerted on the box by the surface.

b Given that the box is about to slip, find an expression, in terms of X and θ, for the coefficient of friction between the surface and the box.

2 A small ring of mass $0.2\,\text{kg}$ is threaded on a rough rod which is fixed vertically. The ring is in equilibrium, acted on by a force of magnitude $10\,\text{N}$ pulling upwards at $45°$ to the vertical.

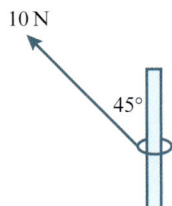

a Show that the frictional force acting on the ring has magnitude $5.07\,\text{N}$, correct to 3 significant figures.

b The ring is on the point of sliding down the rod. Find the coefficient of friction between the ring and the rod.

3 A particle is projected with speed $8\,\text{m}\,\text{s}^{-1}$ across a rough horizontal surface and comes to rest $10\,\text{m}$ from its starting point.

a Calculate the coefficient of friction between the particle and the surface.

b A second, identical particle is projected across the same surface and comes to rest $20\,\text{m}$ from its starting point. Determine its initial speed.

 4 A box of mass $5\,\text{kg}$ is placed on a rough slope inclined at an angle of $20°$ to the horizontal. It is released from rest and slides down the slope.

a Draw a diagram showing the forces acting on the box.

The slope is $1.5\,\text{m}$ long. The box takes 4.8 seconds to reach the bottom of the slope.

b Find the acceleration of the box.

c Find the coefficient of friction between the box and the slope.

d State an assumption that you have made about the forces acting on the box.

5 A block of weight 20 N is at rest on a plane which is inclined
to the horizontal at angle α, where $\sin \alpha = 0.28$. The coefficient
of friction between the block and the plane is μ. A force of
magnitude 7.2 N acting parallel to a line of greatest slope is
applied to the block. When the force acts up the plane, the
block remains at rest.

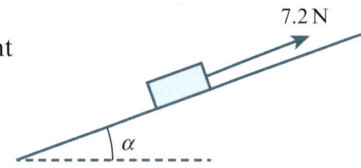

a Find the minimum possible value of μ.

When the force acts down the plane the block is in limiting
equilibrium.

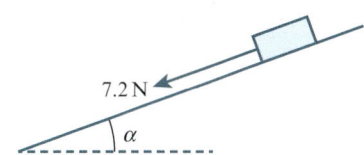

b Find the value of μ.

P **6** A particle is projected with speed $5\,\mathrm{m\,s^{-1}}$ down the line of steepest slope of an inclined rough
plane. The coefficient of friction between the particle and the plane is 1.

It takes the same time T seconds for a particle of mass 2 kg to travel 1 metre from its start
position if the plane is inclined at angle 2θ to the horizontal as it takes for a particle of
mass 1 kg to travel 1 metre from its start position if the plane is inclined at angle θ to the
horizontal.

Show that $(4\cos\theta - 1)(\cos\theta - \sin\theta) = 2$.

49

PS **7** Two right-angled triangular prisms of equal height, with angle
of greatest slope 30° and 45° respectively, are positioned as
shown, with a smooth peg P between the two highest points.

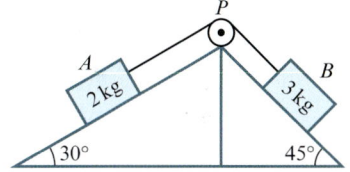

Block A, with mass 2 kg, is placed on the 30° slope and
block B, with mass 3 kg, is placed on the 45° slope. The two
blocks are connected by a light, inextensible string which runs
parallel to the line of greatest slope of each prism and passes over the smooth peg.

The coefficient of friction between block A and the 30° slope surface is μ and the coefficient
of friction between block B and the 45° slope surface is 2μ.

At time $t = 0$, block A is projected down the 30° slope with speed $9\,\mathrm{m\,s^{-1}}$.

a Calculate the acceleration of block A in terms of μ.

It is determined that $\mu = 0.1$.

b Calculate whether the blocks will return to their original positions, and if so, the time at
which this will occur.

Chapter 5
Connected particles

- Use Newton's third law for objects that are in contact.
- Calculate the motion or equilibrium of objects connected by rods.
- Calculate the motion or equilibrium of objects connected by strings.
- Calculate the motion or equilibrium of objects that are moving in lifts (elevators).

 TIP

Unless otherwise stated, g should be taken as $10\,\mathrm{m\,s^{-2}}$.

5.1 Newton's third law

WORKED EXAMPLE 5.1

A boy puts two boxes, one on top of the other, on a low trolley and pushes the trolley across horizontal ground. The boxes do not fall off the trolley. Describe and state the direction of each force that acts on:

a   the upper box

b   the lower box.

Answer

a   Weight of box (vertically downwards).

Contact force (vertically upwards) from contact with the lower box.

The lower box pushes upwards on the upper box.

Friction force (horizontally in the direction of travel) from contact with the lower box.

The upper box moves forwards so there must be a resultant force in the direction of travel.

b   Weight of box (vertically downwards).

Contact force (vertically downwards) from contact with the upper box.

The upper box pushes down on the lower box (with an equal and opposite contact force).

Friction force (horizontally opposing the direction of travel) from contact with upper box.

The friction forces between the boxes are equal and opposite.

Contact force (vertically upwards) from contact with the trolley.

The trolley pushes upwards on the lower box.

Friction force (horizontally in the direction of travel) from contact with the trolley.

The lower box moves forwards so there must be a resultant force in the direction of travel.

1 A crate of weight 80 N is stacked on top of a crate of weight 100 N, on horizontal ground as shown. Make separate sketches showing the forces acting on the upper and lower crates. Indicate the magnitudes of the forces in your sketches.

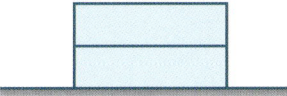

2 A birdbath consists of a concrete pillar of weight 1000 N with a concrete bowl of weight 200 N on top, as shown in the diagram. State the magnitude and direction of the force exerted by:

a the pillar on the bowl

b the bowl on the pillar

c the ground on the pillar.

3 A sack is in contact with both the base and the vertical back of an excavator scoop, as shown in the diagram. The excavator is moving forwards at constant speed in a straight line. Make separate sketches showing the forces acting on:

a the sack

b the scoop.

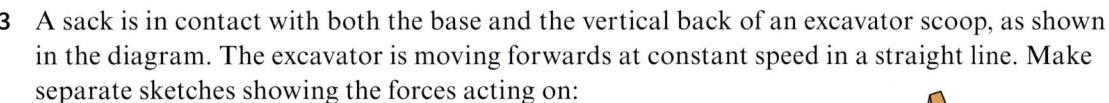

4 A reckless truck-driver loads two identical untethered crates stacked one upon the other as shown in the diagram. No sliding takes place. Make separate sketches to show the forces acting on each crate when the truck travels on a horizontal straight road:

a with constant speed

b while accelerating

c while decelerating.

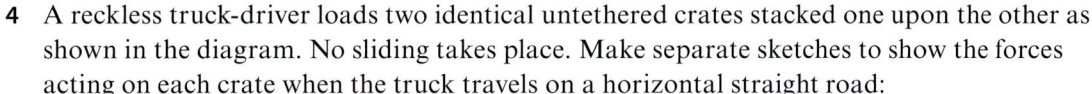

Each crate has mass 250 kg. When the acceleration of the truck is $1.5\,\mathrm{m\,s^{-2}}$, calculate the magnitude of the frictional force exerted on:

d the upper crate by the lower

e the lower crate by the deck of the truck.

5 In an archaeological reconstruction, a group of students try to drag an ancient warship along a beach. This requires a horizontal force of 5000 N. The average weight of a student is 600 N, and the coefficient of friction between each student's feet and the sand is 0.2. How many students are needed?

5.2 Objects connected by rods

TIP

Newton's second law can be applied in a connected system to the entire system or to the individual components of the system.

WORKED EXAMPLE 5.2

A car of mass 1800 kg tows a trailer of mass 2000 kg along a level road. The driving force from the engine is 600 N. The resistance on the car is 50 N and on the trailer is 100 N.

a Find the acceleration of the car.

b Find the tension in the towbar.

Answer

a

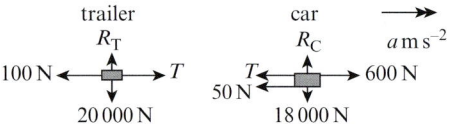

Draw a force diagram.

Newton's second law for the system:

$$600 - 50 - 100 = 3800\,a$$
$$a = 0.118\,\mathrm{m\,s^{-2}}$$

Resolve horizontally in the direction of motion on the whole system.

Solve to find the acceleration.

b Newton's second law for the trailer:

$$T - 100 = 2000\,a$$
$$T = 337\,\mathrm{N}$$

Resolve horizontally in the direction of motion on the trailer only. (Alternatively resolve for the car only.)

Substitute in the acceleration value already found.

EXERCISE 5B

1 A car of mass 1200 kg, towing a caravan of mass 800 kg, is travelling along a motorway at a constant speed of $20\,\mathrm{m\,s^{-1}}$. There are air resistance forces on the car and the caravan, of magnitude 100 N and 400 N respectively. Calculate the magnitude of the force on the caravan from the towbar, and the driving force on the car.

The car brakes suddenly, and begins to decelerate at a rate of $1.5\,\mathrm{m\,s^{-2}}$. Calculate the force on the car from the towbar. What effect will the driver notice?

2   A car of mass 750 kg is towing a trailer of mass 350 kg. The driving force on the car has magnitude 15 kN. The resistance forces on the car and the trailer are 600 N and 400 N, respectively. Find:

   a   the acceleration of the car and the trailer

   b   the tension in the tow bar.

3   A car of mass 1200 kg is towing a trailer of mass 400 kg using a light towbar. The resistance forces acting on the car and the trailer are 500 N and 300 N, respectively. The car starts to brake and decelerates at $1.2\,\mathrm{m\,s^{-2}}$. Find the magnitude of the braking force.

4   A train is made up of an engine of mass 4500 kg and two carriages of mass 2500 kg each. The train is accelerating at $0.9\,\mathrm{m\,s^{-2}}$. The resistance force acting on the engine is 1200 N and the resistance force acting on each carriage is 500 N. Find:

   a   the driving force on the engine   b   the tension in each coupling.

5   A train consists of an engine of mass 4000 kg and two carriages of mass 2000 kg each. The train is decelerating at $2\,\mathrm{m\,s^{-2}}$. The resistance forces are: 9000 N on the engine, 6000 N on the first carriage and 3000 N on the second carriage.

   a   Determine whether the engine is driving or braking.

   b   Find the magnitude of the force in each coupling, stating whether it is a tension or a thrust.

6   A car of mass 1000 kg is towing a trailer of mass 250 kg along a straight road. There are constant resistances to the motion of the car and the trailer of magnitude 150 N and 50 N respectively. The driving force on the car has magnitude 800 N. Calculate the acceleration of the car and the trailer, and the tension in the towbar, when:

   a   the road is horizontal

   b   the road is inclined at $\sin^{-1} 0.04$ to the horizontal and the car is travelling uphill.

7   When a car of mass 1350 kg tows a trailer of mass 250 kg along a horizontal straight road, the resistive forces on the car and trailer have magnitude 200 N and 50 N respectively. Find the magnitude of the driving force on the car when the car and trailer are travelling at constant speed, and state the tension in the towbar in this case.

   Find the acceleration or deceleration of the car and trailer, and the tension in the towbar, when the driving force exerted by the car has magnitude:

   a   330 N   b   170 N   c   zero.

P   8   A car of mass M pulls a trailer of mass m down a straight hill which is inclined at angle $\alpha°$ to the horizontal. Resistive forces of magnitudes P and Q act on the car and the trailer respectively, and the driving force on the car is F. Find an expression for the acceleration, a, of the car and trailer, in terms of F, P, Q, M, m and α.

   Show that the tension in the towbar is independent of α.

   In the case when $F = P + Q$, show that the acceleration is $g \sin \alpha$ and that the tension in the towbar is Q.

5.3 Objects connected by strings

WORKED EXAMPLE 5.3

A box of mass 5 kg sits on a rough horizontal table that is 70 cm high. The coefficient of friction between the table and the box is 0.25. Light inextensible strings are attached to the box, one passing over a small smooth pulley at the left edge of the table and the other passing over a small smooth pulley at the right edge of the table. The free ends of the strings hang vertically. A box of mass 1 kg hangs from the end of the string on the left side of the table and a box of mass 4 kg hangs from the end of the string on the right side of the table.

The system is released from rest and the boxes move without hitting any of the pulleys. How long does it take for the box on the right side to descend to the floor?

Answer

Resolving vertically for the box on the table: | The box on the table does not move vertically.

$R = 50\,\text{N}$

Friction is limiting

$F = 0.25 \times 50 = 12.5\,\text{N}$ | The box is moving so $F = \mu R$.

Newton's second law for each box: | Resolve in the direction of motion for each box.

$40 - T_1 = 4a$ | Right (4 kg) box.

$T_1 - T_2 - 12.5 = 5a$ | (5 kg) box on table.

$T_2 - 10 = a$ | Left (1 kg) box.

Sum to eliminate T_1 and T_2: | Solve the equations simultaneously by adding all three.

$17.5 = 10a$

$a = 1.75\,\text{m s}^{-2}$ | Solve to find the acceleration of each of the three boxes.

Constant acceleration: | State the known values.

$s = 0.70,\ u = 0,\ a = 1.75$ | Box descends 70 cm to floor.

$s = ut + \frac{1}{2}at^2$ | Use the most suitable equation of motion to find the time taken for the 4 kg box to descend to the floor.

$t = 0.894$

It takes 0.894 seconds for the box to reach the floor.

EXERCISE 5C

1   Particle P of mass 4.5 kg is being pulled by a light inextensible string. Another light inextensible string is attached to the other side of P and particle Q, of mass 6 kg, is attached to the other end of this string. The particles move with acceleration 2.1 m s^{-2} in a straight line on a smooth horizontal table. Find the tension in each string.

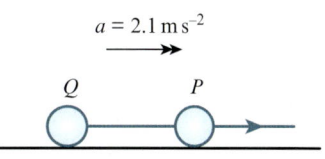

2   Two identical boxes, each of mass 10 kg, are connected by a light inextensible cable. One box is pushed away from the other with a force of 75 N. The boxes move in a straight line at constant velocity on a rough horizontal table. Find the magnitude of the friction force between each box and the table.

3   A box of mass 24 kg is pulled across a rough horizontal floor with a force of F N. The friction force between this box and the floor is 80 N. Another box, of mass 15 kg, is attached to the first box by a light inextensible string. The friction force between the second box and the floor is 50 N.

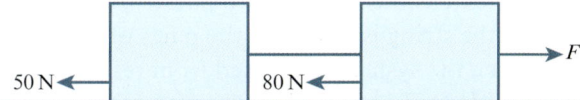

   a   The string connecting the two boxes will break if the tension exceeds 120 N. Find the largest possible value of F.

   b   The string breaks when the boxes are moving with a speed of 2.4 m s^{-1}. Assuming the two boxes do not collide, how long does it take for the second box to stop?

4   A crate of mass 35 kg is suspended by a light inextensible cable. Another crate, of mass 50 kg, is attached to the bottom of the first crate by another light inextensible cable. Find the tensions in the two cables when the crates are:

   a   being raised with acceleration 0.8 m s^{-2}

   b   being lowered at constant speed.

5   Two particles of masses 5 kg and 7 kg are connected by a light inextensible string that passes over a fixed smooth pulley. The system is released from rest with both ends of the string vertical and taut. Find the acceleration of each particle and the tension in the string.

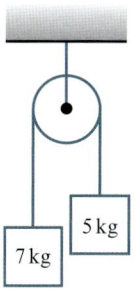

6 Box A, of mass 15 kg, rests on a rough horizontal table. It is connected to one end of a light inextensible string which passes over a fixed smooth peg. Box B, of mass 6 kg, is attached to the other end of the string.

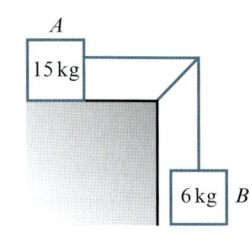

   a Given that the system is in equilibrium, find the magnitude of the frictional force between box A and the table.

   b Given instead that the contact between box A and the table is smooth, find the acceleration of the system and the tension in the string.

7 A particle of mass 3 kg is attached to one end of each of two strings, s_1 and s_2. A particle of mass 2.5 kg is attached to the other end of s_1, and a particle of mass 1 kg is attached to the other end of s_2. The particles are held in the positions shown, with the strings taut and s_1 passing over a smooth fixed peg. The system is released from rest. Find the acceleration of the particles, and the tensions in s_1 and s_2.

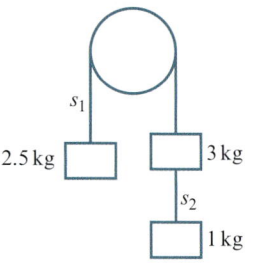

 8 Two particles are connected by a light inextensible string which passes over a smooth fixed peg. The heavier particle is held so that the string is taut, and the parts of the string not in contact with the pulley are vertical. When the system is released from rest the particles have an acceleration of $\frac{1}{2}g$. Find the ratio of the masses of the particles.

9 Two particles have masses m and km, with $k > 1$. The particles are connected by a light inextensible string passing over a smooth pulley. The system is released from rest and the particles move with acceleration $\frac{2}{3}g$. Find the value of k.

10 A particle of mass 4 kg is attached by light inextensible strings to two other particles of masses 7 kg and 5 kg. The string connecting the 4 kg particle to the 7 kg particle passes over a smooth pulley, as shown in the diagram. The particle of mass 7 kg is attached to the floor by another light inextensible string.

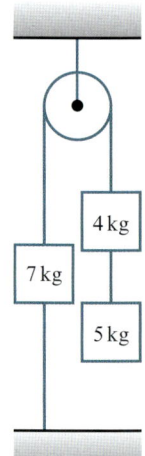

   a Given that the system is in equilibrium, find the tension in the string connecting the 7 kg particle to the floor.

   b This string is now removed. Find the acceleration of the system.

5.4 Objects in moving lifts

WORKED EXAMPLE 5.4

A lift can carry a maximum of 5 persons of average mass 80 kg. The maximum acceleration of the empty lift when it travels upwards is $1\,\mathrm{m\,s^{-2}}$. By considering the lift travelling upwards when empty and when full:

a find the mass of the lift

b find the maximum tension in the lift cable

c comment on the calculated values.

Answer

a

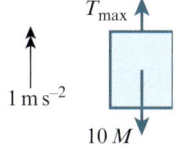

Draw a force diagram.

Newton's second law for the empty lift travelling upwards at maximum acceleration:

$$T_{\max} - 10M = M$$

Resolve in the direction of motion.

so $\quad T_{\max} = 11M$

Find the maximum tension in terms of the mass of the lift.

$T_{\max}$

5 persons of average mass 80 kg have total mass 400 kg.

$a\,\mathrm{m\,s^{-2}}$

$10(M + 400)$

Draw a new force diagram.

Newton's second law for the full lift travelling upwards at acceleration $a\,\mathrm{m\,s^{-2}}$:

$$T_{\max} - 10(M + 400) = (M + 400)a$$

Resolve in the direction of motion.

$a > 0$ so $11M - 10(M + 400) > 0$

$M > 4000$

The mass of the lift is at least 4000 kg.

Replace the maximum tension with the expression found earlier, and consider that the acceleration must be positive.

Solve the inequality for the mass of the lift.

b When $M = 4000\,\mathrm{kg}$, $T_{\max} = 11M = 44\,000\,\mathrm{N}$

Use the expression for the maximum tension found earlier.

c   The value of a will need to be more than just greater than 0, perhaps 0.1, and there needs to be some spare capacity in case some heavy people get into the lift, so the figure 400 kg should be increased, perhaps to 500 kg.

$$T_{max} - 10(M + 500) = (M + 500)a$$

| Resolve in the direction of motion for a full lift with a mass of 500 kg for the load in the lift. |

$$11M - 10(M + 500) = (M + 500)a$$

| Replace the maximum tension with the expression found earlier. |

$$M(1 - a) = (500)(10 + a)$$

| Simplify the equation. |

If $a = 0.1$ then $M = 5611$ kg and $T_{max} = 61\ 700$ N

So the actual values will be greater than the values calculated in parts **a** and **b**.

| Consider an acceleration of 0.1 to see how this affects the mass of the lift and the maximum tension in the lift cable. |

EXERCISE 5D

1   A crate of mass 120 kg lies on the horizontal floor of a lift. The lift accelerates upwards at $0.4 \, \text{m s}^{-2}$. Find the magnitude of the normal reaction force between the crate and the floor of the lift.

2   A basket of mass 750 grams is attached to a light inextensible rope and is being lowered at a constant speed. A box of mass 120 grams rests at the bottom of the basket. Find the magnitude of the normal contact force between the box and the basket.

3   A child of mass 35 kg stands in a lift of mass 500 kg. The lift is suspended by a light inextensible cable and accelerates upwards at $0.6 \, \text{m s}^{-2}$. Find:

   a   the tension in the cable

   b   the magnitude of the normal reaction force between the child's feet and the floor of the lift.

4   A person of mass 85 kg stands in a lift of mass 360 kg. The lift is suspended by a light inextensible cable. Find the magnitude of the normal reaction force between the person's feet and the floor of the lift when the lift is:

   a   moving downwards and decelerating at $4.2 \, \text{m s}^{-2}$

   b   moving upwards at a constant speed.

5   A lift of mass 520 kg is supported by a steel rod attached to its bottom. The rod can withstand a maximum thrust of 15400 N. The lift can accelerate at $2.5 \, \text{m s}^{-2}$ and decelerate at $7.8 \, \text{m s}^{-2}$. Find the maximum allowed load in the lift.

6 A woman of mass 63 kg stands in a lift of mass 486 kg. The lift is supported by a cable and moves with acceleration $2.2\,\text{m s}^{-2}$. The magnitude of the normal reaction force between the woman's feet and the floor of the lift is 765 N.

    a Is the lift going up or down?
    b Find the tension in the cable.

END-OF-CHAPTER REVIEW EXERCISE 5

 1 Particles P and Q, of masses 1 kg and 0.6 kg respectively, are attached to the ends of a light inextensible string which passes over a smooth fixed peg. The particles are held at rest with both hanging at a height of 0.9 m above the floor.

The system is released from rest. Find:

    a the time until P reaches the floor

    b the time between P reaching the ground and P leaving the floor again, assuming that P does not bounce.

2 Particles A and B, of masses m kg and $2m$ kg respectively, are connected by a light inextensible string. The string passes over a smooth pulley at the edge of a table. Particle A hangs freely and particle B is sitting on the table.

    a The system is in limiting equilibrium. Find the coefficient of friction between the table and particle B.

A horizontal force of magnitude F N is then applied to particle B, acting away from the pulley.

    b The system is in limiting equilibrium. Find the value of F.

3 Particles P and Q, of masses M kg and m kg respectively where $M > m$, are attached to the ends of a light inextensible string. The string passes over a small smooth pulley which is fixed at the point where two rough planes meet. The coefficient of friction for each plane is 0.1. The planes each make an angle θ with the ground.

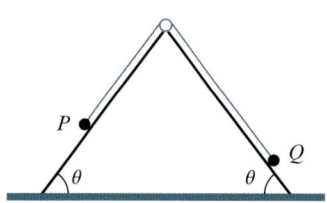

The particles are released from rest.

    a Find, in terms of M, m and θ, the tension in the string while both particles are moving.

Particle P reaches the ground after 2 seconds. The speed of particle P when it reaches the ground is $v\,\text{m s}^{-1}$.

    b Find v in terms of M, m and θ.

Particle P is immediately brought to rest and particle Q continues to moves up the slope.

In the case when $\theta = 60°$, $M = 0.3$ and $m = 0.1$, particle Q just reaches the pulley.

    c Find the height of the pulley above the ground.

59

4   A crate of mass 80 kg lies on a horizontal platform. The platform is being raised and decelerates at $2.6\,\mathrm{m\,s^{-2}}$. Find the magnitude of the normal reaction force between the crate and the platform.

5   A book of mass 310 g lies on a rough horizontal table. A light inextensible string is attached to the book. The string passes over a smooth pulley fixed at the edge of the table. A ball of mass 120 g is attached to the other end of the string.

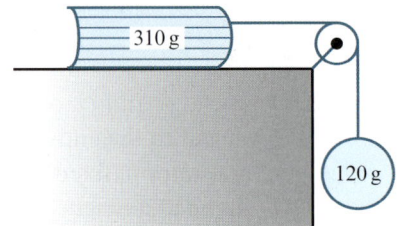

The system is in equilibrium with the string taut. Find the magnitude of the friction force between the book and the table.

6   Machinery of mass 300 kg is placed on the floor of a lift of mass 450 kg. The magnitudes of the tension in the cable holding the lift, and the normal contact force between the machinery and the lift floor, are T newtons and R newtons respectively. By considering the forces acting on the machinery, find the value of R when the lift is moving upwards with a deceleration of $2.2\,\mathrm{m\,s^{-2}}$. By considering the motion of the machinery and the lift as one body, find the value of T for the same deceleration.

7   A car is pulling a trailer using a light rigid tow bar. The mass of the car is 1200 kg and the mass of the trailer is 350 kg. Assume that any resistances to motion can be ignored.

   a   The car is moving with a speed of $9.2\,\mathrm{m\,s^{-1}}$ when it starts to accelerate at $1.8\,\mathrm{m\,s^{-2}}$. Find the driving force acting on the car.

   b   The car continues to accelerate uniformly for 4 seconds. It then starts to brake, with uniform deceleration, until it comes to rest. During the braking phase, it travels 26 m.

   Find the magnitude of the thrust in the towbar during the braking phase.

8   A small ring of mass 0.1 kg is threaded on a fixed vertical rod. A string AB, of length 1.5 m, is attached to the ring at A. A second string BC, of length 2 m, is attached to the rod at a point C. The point C is vertically above A and each string is light and inextensible.

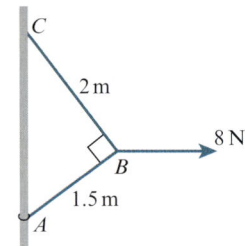

   A horizontal force of magnitude 8 N is applied at the point B. The system is in equilibrium with the strings taut and AB perpendicular to BC.

   a   Find the tension in each string.

   The coefficient of friction between the rod and the ring is μ. The ring is on the point of slipping.

   b   Find the value of μ to 3 s.f.

9 A person of mass 75 kg stands in a lift of mass 450 kg. The lift is suspended by a light inextensible cable and moving downwards.

a The lift is decelerating at $5.2\,\mathrm{m\,s^{-2}}$. Find the normal reaction force between the person's feet and the floor of the lift.

b Given instead that the normal reaction force between the person's feet and the floor of the lift is 577.5 N:

    i find the magnitude and direction of the acceleration of the lift

    ii calculate the tension in the cable.

10 Two skaters stand on ice 5 m from each other, holding onto the ends of a light inextensible rope. They pull at the rope with a constant force and come together in 1.5 seconds. Any friction can be ignored. Given that the mass of the first skater is 62 kg, and that he moves with acceleration $1.8\,\mathrm{m\,s^{-2}}$, find the mass of the second skater.

11 Box A of mass 6 kg is held at rest at one end of a rough horizontal table. The box is attached to one end of a light inextensible string which passes over a smooth pulley fixed to the other end of the table. The length of that part of the string extending from A to the pulley is 3 m. Box B, of mass 2.5 kg, is attached to the other end of the string and hangs 1.2 m above the ground. The system is released from rest and moves with acceleration $0.3\,\mathrm{m\,s^{-2}}$.

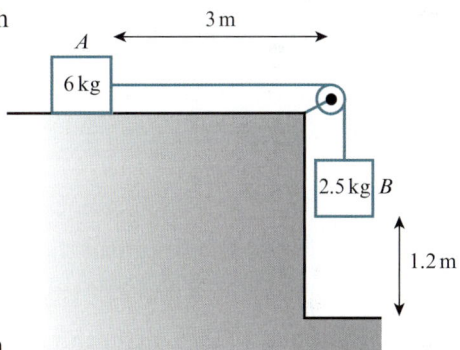

a Find the magnitude of the frictional force between box A and the table.

b Box B reaches the floor and remains at rest. The magnitude of the frictional force between box A and the table remains unchanged. Will box A reach the pulley?

12 Particles P and Q, of masses 3 kg and 5 kg, are connected by a light inextensible string. The string passes over a smooth pulley and the particles hang in the vertical plane with Q 2.5 m above the ground.

At time $t = 0$ the system is released from rest with the string taut.

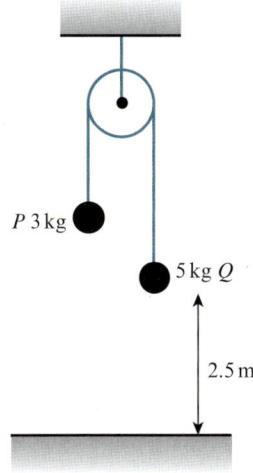

a Find the time required for Q to hit the ground.

Once Q is on the ground, P continues to move. Assume that in subsequent motion, neither particle reaches the pulley.

b Find the greatest height of P above its start point.

c Find the time when the string becomes taut again.

Chapter 6
General motion in a straight line

- Use differentiation to calculate velocity when displacement is given as a function of time.
- Use differentiation to calculate acceleration when velocity is given as a function of time.
- Use integration to find displacement when velocity is given as a function of time.
- Use integration to find velocity when acceleration is given as a function of time.

6.1 Velocity as the derivative of displacement with respect to time

TIP

Remember $v = \dfrac{\mathrm{d}s}{\mathrm{d}t}$.

WORKED EXAMPLE 6.1

A particle moves along the x-axis for 5 seconds. The position of the particle, in metres, at time t seconds after the start is given by:

$$x = \begin{cases} -3t^2 + 8t^{1.5} - 3 & 0 \leqslant t \leqslant 4 \\ A + t - Bt^2 & 4 \leqslant t \leqslant 5 \end{cases}$$

a   What is the position at the start?

b   What is the initial velocity?

There is no sudden change in the velocity when $t = 4$.

c   Find the values of A and B.

d   What are the positions when the particle is stationary?

e   What is the total distance travelled?

Answer

a $x = -3$ ⋯⋯⋯⋯⋯⋯⋯⋯⋯⋯⋯ Put $t = 0$.

b $v = -6t + 12t^{0.5}$ ⋯⋯⋯⋯⋯⋯ Use $v = \dfrac{\mathrm{d}s}{\mathrm{d}t}$
  $t = 0$ gives initial velocity $= 0 \,\mathrm{m\,s^{-1}}$

c $A + 4 - 16B = 13$ ⋯⋯⋯⋯⋯⋯ $t = 4$ gives $x = 13$.
  $\dfrac{\mathrm{d}x}{\mathrm{d}t} = 1 - 8B = 0$ $t = 4$ gives $v = 0$.

  so $A = 11$ and $B = 0.125$ ⋯⋯⋯ Solve to find B, then substitute in the first equation to find A.

d Stationary when $v = 0\,\text{m s}^{-1}$ .... | Stationary means the particle is not moving, so the velocity is zero.

$0 \leqslant t \leqslant 4: -6t + 12t^{0.5} = 0$ ....

$6t^{0.5}(2 - t^{0.5}) = 0$ .... | Set the first velocity expression equal to zero.

$t = 0 \text{ or } 4$

$4 \leqslant t \leqslant 5: 1 - 0.25t = 0$ .... | Factorise to find the time.

$t = 4$

Stationary when $t = 0$ or $t = 4$ | Set the second velocity expression equal to zero.

Positions are $x = -3$, $x = 13$ .... | Substitute the time values into the expressions for displacement to find the positions when the particle is stationary, or use the values found in parts **a** and **c**.

e $t = 5$ gives $x = 12.875$ .... | Starts at $x = -3$ and moves to 13 then instantaneously at rest and then travels to $x = 12.875$.

Total distance $= 16.125\,\text{metres}$

EXERCISE 6A

1 A particle moves along a straight line. The displacement, $s\,\text{m}$, from a fixed point at time $t\,\text{s}$ is given by:

$$s = 5 + 9t - 3.3t^2 + 0.4t^3 \text{ where } 0 \leqslant t \leqslant 4$$

Find the times at which the particle is stationary.

2 A particle moves in a straight line. Its displacement, $x\,\text{m}$, from a fixed point at time $t\,\text{s}$ is given by:

$$x(t) = 2t^3 - 5t$$

a Find the velocity when $t = 2$.

b Find the average velocity between $t = 0$ and $t = 4$.

3 Given that $x = 120 - 15t - 6t^2 + t^3$ where x is measured in metres and t in seconds, find the time when the velocity is zero. Find the displacement at this instant.

4 Given that displacement in metres is $x = 2 + 48t - t^3$ for time $t\,\text{s}$, find the displacement when the velocity is zero.

5 Two ants move around so that the distance, $d\,\text{cm}$, between them at time t seconds is given by:

$$d = 20 - 4t^{0.5} + t$$

What is the minimum separation between the ants?

M **6** A spider runs forwards and backwards along a pipe. The distance, $s\,\text{cm}$, of the spider from the end of the pipe at time t seconds is given by:

$$s = t^3 - 6t^2 + 9t + 10$$

The spider gets to the end of the pipe after 5 seconds.

    a How long is the pipe?

    b How fast is the spider moving when it gets to the end of the pipe?

    c What is the average speed of the spider?

M **7** The height, in metres, of the tide at a certain place on a certain day, at time t hours after midnight, is given by:

$$h = 3 + 2\cos(0.55t)$$

where angles are measured in radians.

    a What is the height of the high tide?

    b What is the time between successive high tides? Give your answer to the nearest minute.

    c When is the tide increasing most rapidly?

6.2 Acceleration as the derivative of velocity with respect to time

WORKED EXAMPLE 6.2

A particle moves along a straight line. The velocity, $v\,\text{m s}^{-1}$, of the particle at time t seconds is given by:

$$v(t) = (0.025)(6 - t)(10 - t) \qquad 0 \leqslant t \leqslant 12$$

Calculate the acceleration of the particle when it is instantaneously stationary.

TIP

Remember

$$a = \frac{\mathrm{d}v}{\mathrm{d}t} = \frac{\mathrm{d}^2 s}{\mathrm{d}t^2}.$$

Answer

$v = 0$ when $t = 6$ or 10 Instantaneously stationary when $v = 0$.

$v = 0.025(t^2 - 16t + 60)$ Expand the double brackets to make differentiation easier.

$a = \dfrac{\mathrm{d}v}{\mathrm{d}t} = 0.025(2t - 16)$ Differentiate with respect to time to find an expression for acceleration.

$\quad = 0.05(t - 8)$

When $t = 6$, $a = -0.1\,\text{m s}^{-2}$ Substitute the time values already found into the acceleration expression.

When $t = 10$, $a = 0.1\,\text{m s}^{-2}$

EXERCISE 6B

1 Given that $v = t^2 - 12t + 40$ in m s^{-1}, find the velocity when the acceleration is zero.

2 Given that $x = t^3 + 4t + 6$ for t s and x m, find expressions for v and a in terms of t. Find the displacement, velocity and acceleration when $t = 2$.

3 Given that $x = 36 - \dfrac{4}{t}$ in metres for t s, find the velocity and acceleration when $t = 2$.

4 An object moves in a straight line. Its displacement, x m, from point O, t seconds after it passes O is given by $x = 0.1t^3 - 1.2t^2 + 3.5t$.

    a Find the speed and the magnitude of the acceleration of the object 6 seconds after passing O.

    b What is the speed of the object when it first returns to O?

    c Find the first time when the speed is zero, and the object's distance from O at this time.

5 A particle moves in a straight line. Given that $v = 4t(8 - t)$, and that the particle has mass 3 kg:

    a find the maximum velocity

    b sketch the (t, v) graph for $0 \leqslant t \leqslant 8$

    c find the resultant force acting on the particle when $t = 2$.

6 A car is accelerating from rest. At time t seconds after starting, the velocity of the car is $v \, \mathrm{m\,s^{-1}}$, where $v = 6t - \dfrac{1}{2}t^2$, for $0 \leqslant t \leqslant 6$.

    a Find the velocity of the car 6 seconds after starting.

    b Find the acceleration of the car when its velocity is $10 \, \mathrm{m\,s^{-1}}$.

7 A train leaves a station and travels in a straight line. After t seconds the train has travelled a distance x metres, where $x = (320t^3 - 2t^4) \times 10^{-5}$. This formula is valid until the train comes to rest at the next station.

    a Find when the train comes to rest, and hence find the distance between the two stations.

    b Find the acceleration of the train 40 seconds after the journey begins.

    c Find the deceleration of the train just before it stops.

    d Find when the acceleration is zero, and hence find the maximum velocity of the train.

M 8 A flare is launched from a hot-air balloon and moves in a vertical line. At time t seconds, the height of the flare is x metres, where $x = 1664 - 40t - \dfrac{2560}{t}$ for $t \geqslant 5$. The flare is launched when $t = 5$.

    a Find the height and the velocity of the flare immediately after it is launched.

    b Find the acceleration of the flare immediately after it is launched, when its velocity is zero, and when $t = 25$.

    c Find the terminal speed of the flare.

    d Find when the flare reaches the ground.

    e Sketch the (t, v) graph and the (t, x) graph for the motion of the flare.

6.3 Displacement as the integral of velocity with respect to time

> **TIP**
>
> Remember $s = \int v \, dt$.

WORKED EXAMPLE 6.3

The velocity, $v \, \text{m s}^{-1}$, of a car under braking is given as a function of time, t s, by:
$$v = 0.25(6 - t)^2$$

a Find the initial deceleration of the car.

b Find the distance travelled under braking until the car comes to a stop.

Answer

a $\quad a = \dfrac{dv}{dt} = -0.5(6 - t)$ Differentiate the expression for velocity with respect to time to find an expression for acceleration.

$\quad\quad t = 0$ gives $a = -3$ Initially, time is zero. Use this to find the acceleration at this time.

$\quad\quad$ Initial deceleration $= 3 \, \text{m s}^{-2}$ As the acceleration is -3, the deceleration is $+3$.

b $\quad$ Stops when $t = 6$ $v = 0$ when $0.25(6 - t)^2 = 0$, so the time is 6.

$\quad\quad s = 0.25 \displaystyle\int_0^6 (36 - 12t + t^2) \, dt$ Expand the double brackets in the expression for velocity, then integrate with respect to time to find an expression for displacement.

$\quad\quad = 0.25 \left[36t - 6t^2 + \dfrac{1}{3}t^3 \right]_0^6$

$\quad\quad = 18 \, \text{metres}$ Use limits of 0 and 6 for the time to find the distance travelled.

EXERCISE 6C

> **TIP**
>
> Unless you are told otherwise, in work involving calculus the displacement will be measured from the initial position, so s will be 0 when $t = 0$.

1 A particle moves in a straight line. Its velocity, $v \, \text{m s}^{-1}$, at time t seconds is given by $v = 2.4t - 1.5t^2$.

     a Find the acceleration of the particle after 2 seconds.

     b Find the velocity at the point when the acceleration is zero.

     c The displacement of the particle from its initial position is x m. Find an expression for x in terms of t.

2 Given that $v = 3t^2 + 8$ and that the displacement is $4\,\text{m}$ when $t = 0$, find an expression for x in terms of t. Find the displacement and the velocity when $t = 2$.

3 Given that $v = 6\sqrt{t}$ and that the displacement is $30\,\text{m}$ when $t = 4$, find the displacement, velocity and acceleration when $t = 1$.

4 Given that $v = 3t^2 + 4t + 3$ in $\text{m}\,\text{s}^{-1}$, find the distance travelled between $t = 0$ and $t = 2$.

5 Given that $v = 4 + 3\sqrt{t}$ in $\text{m}\,\text{s}^{-1}$, find the distance travelled between $t = 1$ and $t = 4$.

6 Given that $v = \dfrac{1}{t^2} - 2$ in $\text{m}\,\text{s}^{-1}$, find the distance travelled between $t = 1$ and $t = 5$.

7 Given that $v = 3(t - 3)(t - 5)$ in $\text{m}\,\text{s}^{-1}$, find the distance travelled:

a between $t = 0$ and $t = 3$

b between $t = 3$ and $t = 5$

c between $t = 5$ and $t = 6$.

Hence find the total distance travelled between $t = 0$ and $t = 6$. Sketch the (t, v) graph and the (t, x) graph (assume that the displacement is zero when $t = 0$). How far is the particle from its starting point when $t = 6$?

PS 8 A particle moves in a straight line. Its velocity, $v\,\text{m}\,\text{s}^{-1}$, at time t seconds is given by $v(t) = (t + 3)(t - 2)(t - 7)$. Find the average speed of the particle during the first 7 seconds.

PS 9 The velocity of a particle is given by:

$$v(t) = \begin{cases} 5t - \dfrac{1}{2}t^2 & \text{for } 0 \leqslant t \leqslant 5 \\[2mm] 7.5 - \dfrac{1}{2}t & \text{for } 5 < t \leqslant 15 \end{cases}$$

The average speed of the particle during the first T seconds is $4\,\text{m}\,\text{s}^{-1}$. Find the value of T, where $T < 15$.

PS 10 A particle moves in a straight line. Its displacement, $x\,\text{m}$, from point O is given by $x(t) = 24t - 3t^2$, where t is measured in seconds. The average velocity of the particle during the first T seconds is $9\,\text{m}\,\text{s}^{-1}$. Find its average speed during this time.

> **TIP**
>
> The velocity jumps suddenly when $t = 5$. This can happen, for example, in a collision. The displacement still needs to be continuous.

6.4 Velocity as the integral of acceleration with respect to time

> **TIP**
>
> Remember $v = \int a\, dt$.

WORKED EXAMPLE 6.4

The game of boccia is a variant of bowls for Paralympic athletes. In one game a player rolls a ball down a curved slide and onto the target area. The player gives the ball an initial speed of $0.5\,\mathrm{m\,s^{-1}}$. The acceleration, $a\,\mathrm{m\,s^{-2}}$, of the ball as it descends the slide is modelled as a function of time, t s, by:

$$a = 8 - 0.2t$$

The length of the slide is 1.73 metres.

    a Show that the ball takes about 0.6 seconds to reach the end of the slide.

    b Calculate the speed of the ball when it reaches the end of the slide.

Answer

a $v = \int (8 - 0.2t)\,dt$ Integrate acceleration with respect to time to find an expression for velocity.

    $= 8t - 0.1t^2 + c_1$ Use a constant of integration.

    $v = 0.5$ when $t = 0$ so $c_1 = 0.5$ Use initial conditions to find the constant of integration.

    $v = 8t - 0.1t^2 + 0.5$ Substitute in the value of the constant to give an expression for velocity.

    $s = \int (8t - 0.1t^2 + 0.5)\,dt$ Integrate velocity with respect to time to find an expression for displacement.

    $= 4t^2 - 0.0333t^3 + 0.5t + c_2$ Use a constant of integration.

    At start of slide, $s = 0$
    when $t = 0$ so $c_2 = 0$ Use initial conditions to find the constant of integration.

    $s = 4t^2 - 0.0333t^3 + 0.5t$ Substitute in the value of the constant to give an expression for displacement.

    At end of slide, $s = 1.73$

    $4(0.6)^2 - 0.0333(0.6)^3 + (0.5)(0.6) = 1.73$
    to 3 significant figures   Use the value of 0.6 seconds given in the question to show the displacement is 1.73 metres.

    It takes about 0.6 seconds for the ball to reach the end of the slide.

    State the conclusion.

b When $t = 0.6, v = 5.264\,\mathrm{m\,s^{-1}}$ Use the expression found for velocity with a time of 0.6 seconds.

EXERCISE 6D

1   A car starts from rest and moves in a straight line. Its acceleration, $a\,\mathrm{m\,s^{-2}}$, is given by $a(t) = 0.12t^2 - 1.44t + 4.32$.

   a   Find the equations for the car's velocity and its displacement from the starting point.

   b   Find the velocity and the displacement at the point when the acceleration is zero.

2   A particle moves in a straight line. Its displacement from the point P is x metres and its acceleration is $a = (1 - 0.6t)\,\mathrm{m\,s^{-2}}$. The particle is initially $25\,\mathrm{m}$ from P and moving away from P with velocity $7.5\,\mathrm{m\,s^{-1}}$.

   a   Find an expression for the velocity in terms of t.

   b   Find the particle's displacement from P after 10 seconds.

   c   Find the particle's displacement from P at the time when its acceleration is $-2\,\mathrm{m\,s^{-2}}$.

3   A particle moves in a straight line with acceleration $a = (2 - 6t)\,\mathrm{m\,s^{-2}}$, where the time is measured in seconds. When $t = 2$ its velocity is $-8\,\mathrm{m\,s^{-1}}$. Find the average velocity of the particle between $t = 5$ and $t = 8$.

4   Given that $a = 10 - 6t^2$, and that the velocity is $4\,\mathrm{m\,s^{-1}}$ and the displacement is $12\,\mathrm{m}$ when $t = 1$, find expressions for v and x in terms of t. Find the displacement, velocity and acceleration when $t = 0$.

5   Given that $a = 2 - \dfrac{6}{t^3}$, and that the velocity is $6\,\mathrm{m\,s^{-1}}$ and the displacement is zero when $t = 1$, find the displacement, velocity and acceleration when $t = 3$.

6   Given that $a = 3t - 12$, and that the velocity is $30\,\mathrm{m\,s^{-1}}$ and the displacement is $4\,\mathrm{m}$ when $t = 0$, find the displacement when the acceleration is zero.

7   Given that $a = 4t - 1$ and that the velocity is $5\,\mathrm{m\,s^{-1}}$ when $t = 0$, find the distance travelled between $t = 0$ and $t = 3$.

8   A particle moves in a straight line. Given that $a = \dfrac{1}{t^2}$ and that the velocity is $4\,\mathrm{m\,s^{-1}}$ when $t = 1$, find the velocity when $t = 100$. State the terminal speed of the particle.

(P)   9   Given that $a = -\dfrac{4}{t^3}$, and that the velocity is $2\,\mathrm{m\,s^{-1}}$ and the displacement is zero when $t = 1$, find the displacement when $t = 4$. Show that the velocity is always positive but the displacement never exceeds $2\,\mathrm{m}$.

10   A truck, with initial velocity $6\,\mathrm{m\,s^{-1}}$, brakes and comes to rest. At time t seconds after the brakes are applied the acceleration is $a\,\mathrm{m\,s^{-2}}$, where $a = -3t$. This formula applies until the truck stops.

   a   Find the time taken for the truck to stop.

   b   Find the distance travelled by the truck while it is decelerating.

   c   Find the greatest deceleration of the truck.

69

11 A force of $(36 - t^2)$ newtons acts at time t seconds on a particle of mass 2 kg. When $t = 0$ the particle has velocity $2\,\mathrm{m\,s^{-1}}$ in the direction of the force. Find the velocity of the particle when $t = 6$ and find the distance travelled between $t = 0$ and $t = 6$.

12 A train is travelling at $32\,\mathrm{m\,s^{-1}}$. The driver brakes, producing a deceleration of $k\sqrt{16 - t}\,\mathrm{m\,s^{-2}}$ after t seconds. The train comes to rest in 16 seconds. Find k, and find how far the train travels before coming to rest.

END-OF-CHAPTER REVIEW EXERCISE 6

1 Salma runs in a straight line. At time t seconds, her displacement, x metres, from the starting point satisfies the equation $x = 1.8\,t^2 - 0.2t^3$.

   a Show that Salma starts running from rest.

   b Find her maximum velocity.

2 The velocity of a car, $v\,\mathrm{m\,s^{-1}}$ at time t seconds, satisfies:

$$v = \begin{cases} 4t - 0.5t^2 & \text{for } 0 \leqslant t \leqslant 4 \\ 0.4t^2 - 6.4t + 27.2 & \text{for } t > 4 \end{cases}$$

   a Find the acceleration of the car after 5 seconds.

   b Find the car's displacement from the starting point when:

      i $t = 3$ **ii** $t = 10$.

3 A dog runs past a tree when $t = 0$ with the speed of $3.7\,\mathrm{m\,s^{-1}}$. It accelerates for 5 seconds so that its speed satisfies $v = (u + 0.4t)\,\mathrm{m\,s^{-1}}$.

   a Write down the value of u.

   For the next 5 seconds, the dog decelerates and its speed satisfies:

$$v = \left(\frac{142.5}{t^2}\right)\,\mathrm{m\,s^{-1}}$$

   b Find the dog's final speed.

   c Find the dog's final displacement from the tree.

   d After how long is the dog 32 m from the tree?

4 The acceleration of a particle moving in a straight line, $a\,\mathrm{m\,s^{-2}}$, satisfies $a = 0.1(t - 5)^2$ for $0 \leqslant t \leqslant 5$ seconds. The particle is initially at rest.

   a Explain why the velocity of the particle is never negative between $t = 0$ and $t = 5$.

   b Find the average speed of the particle between $t = 0$ and $t = 5$.

   c By sketching the graph of $x(t)$ show that there is a value of t where the instantaneous speed equals the average speed.

5 A particle starts from rest. Its acceleration after t seconds is $\dfrac{1}{(1 + 2t)^3}\,\mathrm{m\,s^{-2}}$. Find its speed and how far it has moved after 2 seconds.

6 A particle moving in a straight line has displacement x metres after t seconds for $0 \leqslant t \leqslant 5$, where $x = t^3 - 12t^2 + 21t + 18$.

   a Find the displacement, velocity and acceleration when the time is zero.

   b Find the time, velocity and acceleration when the displacement is zero.

   c Find the time, displacement and acceleration when the velocity is zero.

   d Find the time, displacement and velocity when the acceleration is zero.

7 A particle moving on the x-axis has displacement x metres from the origin O after t seconds for $0 \leqslant t \leqslant 5$, where $x = t^2(t - 2)(t - 5)$.

   a For what values of t is the particle on the positive side of O?

   b For what values of t is the particle moving towards O?

   c For what values of t is the force on the particle directed towards O?

8 The displacement, x metres, of a particle moving on the x-axis at a time t seconds after it starts to move is given by $x = t^5$ for $0 \leqslant t \leqslant 1$, and by $x = \dfrac{4}{t} - \dfrac{3}{t^k}$ for $t \geqslant 1$.

   a Verify that both formulae give the same value for x when $t = 1$.

   b Find the value of k for which there is no sudden change of velocity when $t = 1$.

   For the rest of the question, take k to have the value you found in part **b**.

   c Show that the particle stays on the same side of O throughout the motion.

   d What is the greatest distance of the particle from O?

   e For what values of t is the force on the particle acting in the positive direction?

9 A particle moves with velocity $v \, \mathrm{m\,s^{-1}}$, where:

$$v(t) = \begin{cases} 0.16t^3 - 0.12t + 10.6 & \text{for } 0 \leqslant t \leqslant 5 \\ 40 - 2t & \text{for } \quad t > 5 \end{cases}$$

   Find the two times when the particle is 200 m away from the starting point.

PS 10 A car is travelling along a road that has a speed limit of 90 km h^{-1}. The speed of cars on the road is monitored via average speed check cameras, which calculate the average speed of a car by measuring how long it takes to travel a specified distance.

The car starts from rest next to one of the cameras.

Its velocity in m s^{-1} is given by $v(t) = \dfrac{1}{5}t(t - 10)^2$,

and it comes to rest after 10 seconds. It stays stationary for T seconds and then starts moving again with a constant acceleration of 3.5 m s^{-2}. The velocity–time graph of the car's motion is shown in the diagram.

The second camera is positioned 300 m away from the first one.

a Find the time the car takes to reach the second camera after it has started from rest the second time.

b Show that the car's speed exceeded $90\,\mathrm{km\,h^{-1}}$ during both stages of motion.

c The cameras did not detect the car breaking the speed limit. Find the smallest possible value of T.

11 Cars can go over speed bumps at $5\,\mathrm{km\,h^{-1}}$. For an average car, the maximum acceleration is $1.4\,\mathrm{m\,s^{-2}}$ and the maximum deceleration is $3.2\,\mathrm{m\,s^{-2}}$. How far apart should the speed bumps be placed to restrict the maximum speed to $30\,\mathrm{km\,h^{-1}}$?

12 A goods train is moving along a straight track. The velocity, $v\,\mathrm{ms^{-1}}$ at time $t\,\mathrm{s}$ after the train starts to move is given by:

$$v = A(t - 0.02t^2) \quad \text{for } 0 \leqslant t \leqslant T$$
$$v = B \qquad\qquad\quad \text{for } T \leqslant t \leqslant 4T$$
$$v = C(t - 0.02t^2) \quad \text{for } 4T \leqslant t \leqslant 5T$$

where A, B, C and T are constants.

The train is at rest when $t = 0$ and when $t = 5T$.

a Find the value of T.

b Find B and C in terms of A.

The train travels 565 metres between the times $t = 0$ and $t = 5T$.

c Find the maximum speed of the train during its journey.

Chapter 7
Momentum

- Calculate the momentum of a moving body or a system of bodies.
- Use the principle of conservation of momentum to solve problems involving the direct impact of two bodies that separate after impact.
- Use the principle of conservation of momentum to solve problems involving the direct impact of two bodies that coalesce on impact.

 TIP

Unless otherwise stated, g should be taken as $10\,\mathrm{m\,s^{-2}}$.

7.1 Momentum

WORKED EXAMPLE 7.1

A ball of mass $0.05\,\mathrm{kg}$ falls 3.2 metres onto hard ground. The ball starts at rest and falls under gravity. Resistance forces are negligible.

  a Calculate the momentum of the ball just before impact with the ground.

  b The ball rebounds to a height of 1.8 metres. Calculate the change in the momentum in the bounce.

TIP

Remember momentum is given by mv where a body of mass $m\,\mathrm{kg}$ is moving with a velocity of $v\,\mathrm{m\,s^{-1}}$.

73

Answer

  a $s = 3.2,\ u = 0,\ a = 10$ Taking down as positive, list the known values with acceleration due to gravity.

    $v^2 = u^2 + 2as$ Choose the equation of motion to find the velocity just before impact with the ground.

    $v = 8$ Substitute in values and calculate the velocity.

    The ball is travelling at $8\,\mathrm{m\,s^{-1}}$ downwards when it hits the ground.

    Momentum $= mv = 0.4\,\mathrm{N\,s}$ downwards   Substitute the values for m and v into the formula for momentum, and remember to give units.

b $s = 1.8$, $v = 0$, $a = -10$ • • • Taking up as positive, list the known values with acceleration due to gravity.

$v^2 = u^2 + 2as$ • • • Choose the equation of motion to find the velocity as the ball leaves the ground.

$u = 6$ • • • • • • • • • • • • • • • Substitute in values and calculate the velocity.

The ball is travelling at $6\,\mathrm{m\,s^{-1}}$ upwards when it leaves the ground.

Momentum $= mv = 0.3\,\mathrm{N\,s}$ • • Substitute the values for m and v into the formula for momentum, and remember to give units.

So change in • • • • • • • • • • Direction has changed from momentum $= (0.3) - (-0.4)$ downwards to upwards. If we $= 0.7\,\mathrm{N\,s}$ upwards use down as positive there is a loss in momentum of $0.7\,\mathrm{N\,s}$.

EXERCISE 7A

1 Calculate the momentum, in newton seconds (N s), for each situation. Change to SI units (kg, m, s) before you start, if necessary.

   a A rocket of mass $15\,\mathrm{kg}$ and velocity $150\,\mathrm{m\,s^{-1}}$

   b A cat of mass $4.5\,\mathrm{kg}$ and velocity $5\,\mathrm{m\,s^{-1}}$

   c A marble of mass 10 grams and velocity $20\,\mathrm{m\,s^{-1}}$

   d A car of mass $1200\,\mathrm{kg}$ and velocity $60\,\mathrm{km\,h^{-1}}$

   e A rhino of mass 1.4 tonnes running at $20\,\mathrm{m\,s^{-1}}$

2 A bullet of mass 5 grams is fired from a gun. Its momentum is $4.5\,\mathrm{N\,s}$ as it leaves the gun. Work out its velocity.

3 A car is moving at a velocity of $15\,\mathrm{m\,s^{-1}}$. Its momentum is $18\,000\,\mathrm{N\,s}$. Find its mass, in metric tonnes.

PS 4 A dinghy of mass $600\,\mathrm{kg}$ is moving at $20\,\mathrm{m\,s^{-1}}$ when a wind starts to blow, directly opposing the motion of the dinghy. The wind exerts a constant force of $16\,\mathrm{N}$ for 30 seconds. Find the percentage decrease in the momentum of the dinghy.

5 A bowling ball of mass $2\,\mathrm{kg}$ is travelling at $7\,\mathrm{m\,s^{-1}}$ when it hits a wall. The ball rebounds directly, leaving the wall with speed $5\,\mathrm{m\,s^{-1}}$. Calculate the change in the momentum of the ball.

6   A motorbike of mass $250\,\text{kg}$ is travelling in a straight line. The speed is slowed from $23\,\text{m s}^{-1}$ to $13\,\text{m s}^{-1}$. Calculate the loss in momentum.

7.2 Collisions and conservation of momentum

 TIP

Momentum is conserved in impacts. The total momentum is constant.

WORKED EXAMPLE 7.2

Particles A, B and C lie in a line on a smooth table, with B between A and C. There is a wall on the other side of C. Particle A has mass $4\,\text{kg}$, B has mass $3\,\text{kg}$ and C has mass $2\,\text{kg}$. Initially B is stationary, A is moving with speed $6\,\text{m s}^{-1}$ towards B, and C is moving with speed $1\,\text{m s}^{-1}$ towards B from the other side.

A hits B and after the collision the velocity of A has reduced to $3\,\text{m s}^{-1}$, in the same direction as before the impact.

B then hits C and they coalesce. This object then hits the wall and leaves the wall with half the speed that it had when it hit the wall. Finally BC hits A and brings A to rest. Find the speed of BC when it hits the wall for the second time.

Answer

First impact

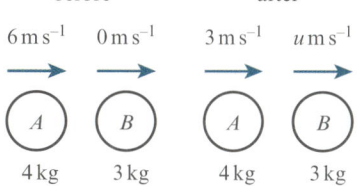

momentum before		momentum after	Momentum is conserved.
$4 \times 6 + 3 \times 0$	$=$	$4 \times 3 + 3u$	
24	$=$	$12 + 3u$	

$u = 4\,\text{m s}^{-1}$ towards C .... Solve the equation to find the velocity of B after the first impact.

Second impact

before   after .... Draw a diagram to summarise the new information.

$4\,\text{m s}^{-1}$ $1\,\text{m s}^{-1}$ $v\,\text{m s}^{-1}$

→   ←   →

(B)   (C)   (BC)

$3\,\text{kg}$ $2\,\text{kg}$ $5\,\text{kg}$ .... The particles coalesce after the second impact, so the combined mass is $5\,\text{kg}$.

75

| momentum before | | momentum after | Momentum is conserved. |

$$3 \times 4 + 2 \times -1 \quad = \quad 5v$$

Remember that momentum is a vector quantity.

$$10 \quad = \quad 5v$$

$v = 2\,\mathrm{m\,s^{-1}}$ towards the wall

Solve the equation to find the velocity of the combined particle after the second impact.

Third impact

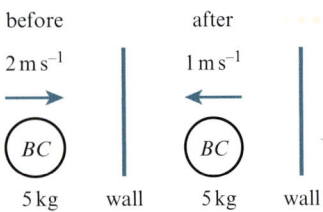

before   after

$2\,\mathrm{m\,s^{-1}}$ $1\,\mathrm{m\,s^{-1}}$

BC   BC

5 kg   wall   5 kg   wall

Draw a diagram to summarise the new information.

BC hits the wall with speed $2\,\mathrm{m\,s^{-1}}$ and leaves the wall with speed $1\,\mathrm{m\,s^{-1}}$ in the opposite direction.

Fourth impact

A comes to rest and BC cannot pass through A so it must reverse its direction of travel.

before   after

$3\,\mathrm{m\,s^{-1}}$ $1\,\mathrm{m\,s^{-1}}$ $0\,\mathrm{m\,s^{-1}}$ $w\,\mathrm{m\,s^{-1}}$

A   BC   A   BC

4 kg   5 kg   4 kg   5 kg

Draw a diagram to summarise the new information.

| momentum before | | momentum after | Momentum is conserved. |

$$4 \times 3 + 5 \times -1 \quad = \quad 4 \times 0 + 5w$$

Remember that momentum is a vector quantity.

$$7 \quad = \quad 5w$$

$w = 1.4\,\mathrm{m\,s^{-1}}$ towards the wall

Solve the equation to find the velocity of the combined particle after the fourth impact.

BC hits the wall for the second time with speed $1.4\,\mathrm{m\,s^{-1}}$

Taking the original direction of travel for A as the positive direction.

EXERCISE 7B

1 A ball, A, of mass $1.5\,\text{kg}$ is travelling at $3\,\text{m}\,\text{s}^{-1}$. It collides with a stationary ball, B, of mass $2\,\text{kg}$. Ball A is brought to rest in the collision. Find the speed of ball B immediately after the collision.

2 A ball, A, of mass $2\,\text{kg}$, moving at $8\,\text{m}\,\text{s}^{-1}$, collides with a ball, B, of mass $2.5\,\text{kg}$, moving towards it at $10\,\text{m}\,\text{s}^{-1}$ in the same straight line. If ball A then moves in a direction opposite to its original path, at a velocity of $6\,\text{m}\,\text{s}^{-1}$, find the final speed of ball B.

3 A cat of mass $3.5\,\text{kg}$ jumps onto a stationary toy train of mass $1.5\,\text{kg}$ that is free to move on a straight horizontal track. The velocity of the cat in the direction of the track immediately before it lands on the train is $20\,\text{m}\,\text{s}^{-1}$. Find the speed of the cat and the train, in the direction of the track, immediately after the cat lands on the train.

4 Particle, A, of mass $0.7\,\text{kg}$, is moving with velocity $8\,\text{m}\,\text{s}^{-1}$ towards particle B, of mass $0.9\,\text{kg}$. Particle B is moving towards A, in the same straight line, with a velocity of $6\,\text{m}\,\text{s}^{-1}$. The particles coalesce. Calculate the magnitude and direction of the velocity of the combined particles.

5 Particle A, of mass $0.8\,\text{kg}$, is moving with velocity $8\,\text{m}\,\text{s}^{-1}$ towards particle B. Particle B is moving towards particle A, in the same straight line, with velocity $6\,\text{m}\,\text{s}^{-1}$. The two particles collide. After the collision the two particles move in opposite directions, both with speed $4\,\text{m}\,\text{s}^{-1}$. Find the mass of particle B.

6 A skateboard of mass $2\,\text{kg}$ is moving at $14\,\text{m}\,\text{s}^{-1}$ when it hits a football of mass 400 grams, which is at rest. Immediately after the collision the skateboard moves at $12\,\text{m}\,\text{s}^{-1}$ in the same straight line and in the same direction. Find the speed of the football immediately after the collision.

PS 7 A particle A, of mass $m\,\text{kg}$, is moving along a straight line with speed $5\,\text{m}\,\text{s}^{-1}$. It collides with another particle, B, of mass $3\,\text{kg}$, moving towards it on the same straight line at a speed of $3\,\text{m}\,\text{s}^{-1}$. After the collision the particles coalesce and move at a speed of $v\,\text{m}\,\text{s}^{-1}$.

   a Find an expression for the two possible values of v, in terms of m.

   b In each case the magnitude of v is $2\,\text{m}\,\text{s}^{-1}$. Find the two possible values of m.

END-OF-CHAPTER REVIEW EXERCISE 7

1   A raindrop of mass $0.005\,\text{kg}$ has momentum of magnitude between $0.01\,\text{N s}$ and $0.02\,\text{N s}$. Calculate the range of speeds of the raindrop.

2   A train of mass $300\,000\,\text{kg}$ is travelling along a straight track. The train speeds up from $4\,\text{m s}^{-1}$ to $6.5\,\text{m s}^{-1}$. Calculate the increase in the momentum of the train, measured in the direction of travel.

3   A toy car of mass $0.3\,\text{kg}$ is at rest when it is hit from behind by another toy car. The second car has mass $0.2\,\text{kg}$ and was moving at speed $1.6\,\text{m s}^{-1}$ immediately before the collision.

     The cars become joined together and continue to move together. Find the common speed of the cars immediately after the collision.

4   A curling stone, of mass $17.5\,\text{kg}$, is moving in a straight line across a horizontal ice rink. The stone is moving at $3.7\,\text{m s}^{-1}$ when it collides with a stationary curling stone. The second stone has mass $m\,\text{kg}$. Immediately after the collision, the stones move together with a common speed of $1.75\,\text{m s}^{-1}$.

     Find m.

5   In a simple model of an impact between a proton and an electron, a proton is moving in a straight line at speed $2.12 \times 10^3\,\text{m s}^{-1}$ when it collides head on with an electron. After the impact the electron has speed $2.09 \times 10^6\,\text{m s}^{-1}$ and the proton is stationary.

     The mass of a proton is 1836 times the mass of an electron.

     Calculate the speed of the electron immediately before the collision.

PS 6   Two identical balls, A and B, travel directly towards one another in a straight line. When they impact ball A is travelling with twice the speed of ball B. After the impact the balls are travelling in the same direction with ball B travelling at twice the speed of ball A. Find the speed of ball B before the impact as a multiple of its speed after the impact.

7   Two balls, X and Y, of masses $0.3\,\text{kg}$ and $0.2\,\text{kg}$ respectively, are thrown into a box that sits on a smooth horizontal floor. When the balls land in the box the horizontal component of the velocity of ball X is $8\,\text{m s}^{-1}$ and the horizontal component of the velocity of ball Y is $9\,\text{m s}^{-1}$ in the opposite direction. The box, with the balls sitting inside it, slides across the floor with speed $0.2\,\text{m s}^{-1}$. Find the mass of the box.

PS 8   Four balls, A, B, C and D, of masses $1\,\text{kg}$, $2\,\text{kg}$, $0.5\,\text{kg}$ and $10\,\text{kg}$ respectively, lie in a line on a smooth table in the order $ABCD$. Initially B, C and D are stationary.

     Ball A is projected towards ball B with speed $0.4\,\text{m s}^{-1}$. After this impact ball B hits ball C and coalesces with it and then BC hits D and comes to rest. At the end of these impacts the velocity of D is $0.05\,\text{m s}^{-1}$ travelling away from BC. Find the final velocity of A.

Chapter 8
Work and energy

- Calculate the work done by a force in moving a body.
- Calculate the kinetic energy and gravitational potential energy of a body.

8.1 Work done by a force

WORKED EXAMPLE 8.1

A constant force of 20 N acting downwards at 30° to the horizontal is used to push a box of mass 10 kg across a rough horizontal floor. The coefficient of friction between the box and the floor is 0.08 and the box is pushed 5 metres across the floor. Calculate the work done on the box:

    a by the push force   **b** against friction.

The box is then pushed by the same force down a slope that is parallel to the force. The coefficient of friction between the box and the slope is 0.08 and the box is pushed 5 metres down the slope. Calculate the work done on the box:

    c by the push force   **d** against friction

    e by the weight of the box.

TIP

Work done by a force F N moving a body a distance d m, at an angle θ to the direction of the force is $W = Fd\cos\theta$.

Answer

a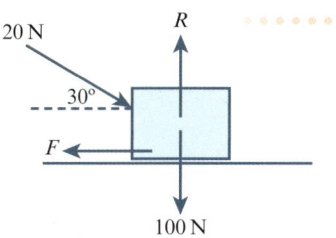

Draw a force diagram.
Weight $= 10 \times 10 = 100$ N.

Work done by the push force
$= 20\cos 30° \times 5$
$= 86.6\,\mathrm{J}$

Box moves horizontally.
Work $= Fd\cos\theta$.

b Resolving vertically:
$R = 100 + 20\sin 30°$
$= 110\,\mathrm{N}$
$F = \mu R = 8.8\,\mathrm{N}$

Resolve vertically to find the normal contact force.

Box is moving so friction is limiting so use $F = \mu R$.

Work done against friction

$= 8.8 \times 5$

$= 44\,\text{J}$

Use $W = Fd$ as the box moves along the line of action of the force.

c

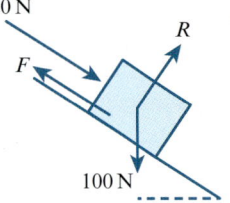

Draw a new force diagram.

Work done by the push force

$= 20 \times 5$

$= 100\,\text{J}$

Box moves down the slope.

d   Resolving perpendicular to slope:

Angle between normal and vertical $= 30°$.

$R = 100 \cos 30° = 86.6\,\text{N}$

Resolve perpendicular to the slope to find the normal contact force.

$F = \mu R = 6.93\,\text{N}$

Box is moving so friction is limiting so use $F = \mu R$.

Work done against friction

$= 6.93 \times 5$

$= 34.6\,\text{J}$

Use $W = Fd$ as the box moves along the line of action of the force.

e   Work done by weight

$= 100 \cos 60° \times 5$

$= 250\,\text{J}$

Angle between slope and vertical $= 60°$, use $W = Fd \cos \theta$.

80

EXERCISE 8A

1   A ball of mass 100 g is dropped from a window. Calculate the work done by gravity as it falls vertically to the ground 6.0 m below.

2   A puck slides 50 metres across an ice rink, against a resistive force of 2.5 N. Calculate the work done against resistance.

3   A cyclist travelling on horizontal ground pedals with a force of 25 N against a headwind of 10 N and a resistance from the road surface of 5.0 N. The cyclist travels 1.2 km. Find:

  a   the work done by the cyclist

  b   the total work done against wind and surface resistances.

4   An object of mass 20 kg is pulled 7 metres at a constant speed across a rough horizontal floor by means of a horizontal rope. The tension in the rope is 100 N. Calculate the work done by the rope. State the work done by the weight of the object and the normal contact force between the object and the floor. State also the work done against resisting forces.

5   A gardener moves a wheelbarrow 30 metres along a level, straight path. The work done by the gardener is 120 J, and the barrow is initially and finally at rest. Calculate the average force resisting the motion.

6   A vehicle engine with driving force 400 N does 50 kJ of work moving the vehicle along a horizontal rough road from A to B. Resistance to motion averages 185 N. Calculate the work done against resistance as the van moves from A to B.

7   A crate is moved at a steady speed in a straight line by means of a towrope. The work done in moving the crate 16 metres is 800 J. Calculate the resolved part of the tension in the rope in the direction of motion.

8   A father pulls his children on a wagon along a level path. The rope by which the wagon is pulled makes an angle of 20° with the horizontal, and has tension 30 N. Calculate the work done in moving the wagon 40 metres at constant speed.

8.2 Kinetic energy

> **TIP**
>
> Remember $KE = \frac{1}{2}mv^2$.

WORKED EXAMPLE 8.2

a   Calculate the kinetic energy of a bag of mass 5 kg in a car moving at $10\,\text{m s}^{-1}$.

The car stops and the bag is dropped over the edge of a cliff. The bag falls from rest; air resistance may be ignored.

b   Calculate the distance that the bag has fallen when it has twice the kinetic energy that it had in part **a**.

Answer

a $KE = \frac{1}{2} \times 5 \times 10^2$ ·········· $KE = \frac{1}{2}mv^2$.

$= 250\,\text{J}$

b   When the bag has fallen h metres: ···· Taking down as positive.

$s = h,\ u = 0,\ a = 10$ ········· State the known values.

$v^2 = u^2 + 2as$ ············· Use a constant acceleration equation to find the square of the velocity in terms of the distance fallen.

$= 20h$

$$KE = \frac{1}{2} \times 5 \times 20\,h$$ ·············· Find the kinetic energy of the bag at this point.

$$= 50\,h$$
$$50\,h = 2 \times 250$$ ·············· Twice the KE from part **a**.
$$h = 10\text{ metres}$$ ·············· Solve the equation to find the distance fallen.

EXERCISE 8B

1 Find the kinetic energy of:

 a a football player of mass 90 kg running at $6\,\mathrm{m\,s^{-1}}$

 b an elephant of mass 6 tonnes charging at $10\,\mathrm{m\,s^{-1}}$

 c a racing car of mass 1.5 tonnes travelling at 300 km per hour

 d a bullet of mass 20 grams moving at $400\,\mathrm{m\,s^{-1}}$

 e a meteorite of mass 20 kg as it enters the Earth's atmosphere at $8\,\mathrm{km\,s^{-1}}$.

2 Calculate the kinetic energy of a cyclist and her bicycle, having a combined mass of 70 kg, travelling at $12\,\mathrm{m\,s^{-1}}$. Give your answer in kJ.

3 Calculate the mass of an athlete who is running at $8.5\,\mathrm{m\,s^{-1}}$, with kinetic energy 3500 J.

4 Calculate the speed of a bus of mass 20 tonnes with kinetic energy 1100 kJ . Give your answer in $\mathrm{km\,h^{-1}}$.

PS 5 A ball of mass 0.8 kg is travelling at speed $15\,\mathrm{m\,s^{-1}}$. The ball slows down and loses 36% of its kinetic energy. Calculate the new speed of the ball.

6 A car of mass 800 kg is being driven along a horizontal straight road. The initial speed of the car is $5\,\mathrm{m\,s^{-1}}$ and the final speed is $25\,\mathrm{m\,s^{-1}}$. Calculate the increase in the kinetic energy of the car.

PS 7 A particle of mass 2 kg falls freely from rest. Calculate the kinetic energy of the particle after it has descended 20 metres.

8.3 Gravitational potential energy

WORKED EXAMPLE 8.3

A ball of mass 0.05 kg is thrown vertically upwards with an initial speed of $5\,\mathrm{m\,s^{-1}}$. The ball rises and then falls back down, finishing 2 metres below where it started. Air resistance may be ignored.

 a Calculate the gain in potential energy between the start and the top.

 b Calculate the loss in potential energy between the start and the finish.

TIP

Remember
$PE = mgh = 10mh.$

Answer

a   From the start to the top: ····· Taking up as positive.

$u = 5, \ v = 0, \ a = -10$ ········ State the known values.

$v^2 = u^2 + 2as$ ············· Use a constant
acceleration equation
The ball rises 1.25 metres.   to find the height
gained by the ball.

Gain in PE $= 0.05 \times 10 \times 1.25$ ·· PE $= mgh$.
$= 0.625 \, \text{J}$

b   The finish is 2 metres below
the start.
Loss in PE $= 0.05 \times 10 \times 2$ ····· Use PE $= mgh$.
$= 1 \, \text{J}$

EXERCISE 8C

1   Calculate the loss of potential energy when a mass of 2 tonnes is lowered through 10 m.

2   A mountaineer of mass 65 kg scales a peak 3.2 km high. Calculate her gain in potential energy.

3   A boy of mass 70 kg climbs a vertical rope. His potential energy increases by 2.45 kJ. Calculate the height that he climbs.

4   A golf ball of mass 0.045 kg is given an initial speed of $30 \, \text{m s}^{-1}$.

    a   Calculate the initial kinetic energy of the ball.

    b   The ball lands 20 metres below where it started. Calculate the loss of potential energy of the ball.

5   A bag of mass m kg is dropped h metres, from rest. Air resistance may be ignored.

    a   Calculate the loss of potential energy.

    b   Calculate the gain in kinetic energy.

6   A crate of mass 80 kg slides down a slope that is inclined at 15° to the horizontal. The crate descends 3 metres down the slope.

    a   Calculate the vertical height that the crate descends.

    b   Calculate the loss of potential energy.

END-OF-CHAPTER REVIEW EXERCISE 8

1 A box is pushed across a rough horizontal floor by a force that is at an angle of 12° below the horizontal.

The box travels at a constant speed of $2 \, \text{m s}^{-1}$ for 3.75 seconds, and the work done against friction is 240 J.

Calculate the tension in the rope.

2 A sledge of mass 100 kg is pulled across a horizontal floor by a constant force of magnitude 120 N acting at an angle θ above the horizontal. The total resistance to motion of the sledge has constant magnitude 40 N.

The sledge is initially at rest. After travelling 30 m it has accelerated to a speed of $5 \, \text{m s}^{-1}$.

Work out:

a the increase in the kinetic energy of the sledge

b the work done against the resistance.

The work done by the force that is pushing the sledge is 2450 J.

c Calculate the angle θ.

3 A parcel is dragged 5 metres across a horizontal floor by means of a horizontal rope. The tension in the rope is 12 N. Calculate the work done by the tension in the rope.

4 A box of mass 25 kg is pushed up a slope that is inclined at 10° to the horizontal. The increase in the potential energy of the box is 217 J.

a Calculate the length of the slope to the nearest metre.

b The resistance to the motion is 22 N. Calculate the work done against the resistance.

5 Water is being pumped from a flooded basement. Each minute 200 litres of water are pumped through a height of 5 metres and ejected at a speed of $1.2 \, \text{m s}^{-1}$. The mass of 200 litres of water is 200 kg.

a Calculate the increase in the potential energy of the 200 litres of water.

b Calculate the increase in the kinetic energy of the 200 litres of water.

6 A circus performer jumps off a swing and lands in the safety net. When she leaves the swing she has speed $4 \, \text{m s}^{-1}$ and she is 2.4 m above the safety net.

a Modelling the performer as a particle of mass 56 kg with no resistance forces acting on her, find:

i the kinetic energy of the performer when she leaves the swing

ii the loss of potential energy for the performer between leaving the swing and reaching the safety net.

b Assume that the potential energy lost is all converted into kinetic energy. Calculate the speed of the performer when she reaches the safety net.

7 Priya and Raj stand side by side behind a safety rail at the top of a cliff. They each hold a pebble of mass 350 grams. Priya throws her pebble vertically upwards with speed $4\,\mathrm{m\,s^{-1}}$. At the same time Raj throws his pebble vertically downwards with speed $4\,\mathrm{m\,s^{-1}}$.

    a Calculate the initial kinetic energy of each pebble.

    b Describe what happens to the kinetic energy of Raj's pebble.

    c Describe what happens to the kinetic energy of Priya's pebble.

8 Finn throws a sack of mass 20 kg from a bridge. The sack falls 4 metres vertically downwards.

    a Calculate the potential energy lost by the sack while it is falling through air.

    b The initial velocity of the sack was $1\,\mathrm{m\,s^{-1}}$ vertically downwards. Assuming that the potential energy lost is all converted into kinetic energy, find the speed of the sack when it has fallen 4 metres.

9 A car of mass 1200 kg is travelling down a hill when the brakes and engine fail. The driver steers into an 'escape road' to stop the car. The escape road is a short horizontal side road with a rough surface and then a big pile of sand. The rough surface is 48 m long and slows the car which will then be stopped by the sand. The car is moving at $14\,\mathrm{m\,s^{-1}}$ when it enters the escape road.

    a Calculate the initial kinetic energy of the car.

    The average resistance from the rough surface is 2000 N.

    b Calculate the work done against resistance while the car travels 48 m over the rough surface.

    c Assuming that the change in the kinetic energy is the same as the work done against resistance, calculate the speed of the car when it gets to the sand.

    d The resistance from the sand is 8000 N. Assuming that the change in the kinetic energy is the same as the work done against resistance, calculate how far the car travels into the sand before it is brought to rest.

 10 A block B lies on a rough horizontal plane. Horizontal forces of magnitudes 30 N and 40 N, making angles of α and β respectively with the x-direction, act on B as shown in the diagram, and B is moving in the x-direction with constant speed. It is given that $\cos\alpha = 0.6$ and $\cos\beta = 0.8$.

    a Find the total work done by the forces shown in the diagram when B has moved a distance of 20 m.

    b Given that the coefficient of friction between the block and the plane is $\dfrac{5}{8}$, find the weight of the block.

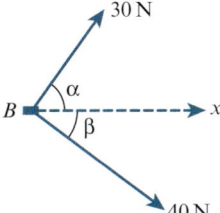

Chapter 9
The work–energy principle and power

- Use the work–energy principle.
- Understand when mechanical energy is conserved.
- Calculate the power of a moving body.
- Use power to calculate the maximum speed of a moving body.

 TIP

Unless otherwise stated, g should be taken as $10\,\mathrm{m\,s^{-2}}$.

9.1 The work–energy principle

WORKED EXAMPLE 9.1

A particle of mass 3 kg slides down a curved track. The particle starts with speed $1\,\mathrm{m\,s^{-1}}$ and descends through 5 m vertically. At the end of the track the particle is moving with speed $4\,\mathrm{m\,s^{-1}}$. Throughout the journey the particle experiences a constant resistance force of 4.25 N. Use the work–energy principle to find the length of the track.

Answer

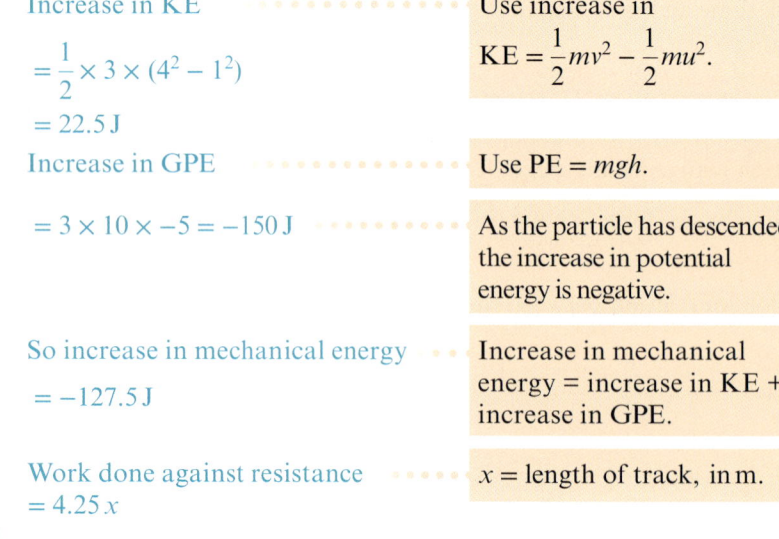

Increase in KE $\cdots\cdots\cdots\cdots$ Use increase in

$= \dfrac{1}{2} \times 3 \times (4^2 - 1^2)$ $\qquad \mathrm{KE} = \dfrac{1}{2}mv^2 - \dfrac{1}{2}mu^2.$

$= 22.5\,\mathrm{J}$

Increase in GPE $\cdots\cdots\cdots\cdots$ Use $\mathrm{PE} = mgh.$

$= 3 \times 10 \times -5 = -150\,\mathrm{J}$ $\cdots\cdots\cdots$ As the particle has descended, the increase in potential energy is negative.

So increase in mechanical energy $\cdots$ Increase in mechanical

$= -127.5\,\mathrm{J}$ $\qquad$ energy = increase in KE + increase in GPE.

Work done against resistance $\cdots\cdots$ x = length of track, in m.
$= 4.25\,x$

 TIP

Remember: increase in mechanical energy = total work done by forces that act to speed the body up − total work done by forces that act to slow the body down.

Increase in mechanical energy =
total work done

$-127.5 = -4.25x$ .... Set up and solve the equation
to find the length of the track.

$x = 30$

The track is 30 metres long.

EXERCISE 9A

1 A stone of mass 0.8 kg is thrown vertically upwards with speed $10 \, \text{m s}^{-1}$. Calculate the initial kinetic energy of the stone, and the height to which it will rise.

2 A child of mass 25 kg slides 15 metres down a water-chute inclined at 30° to the horizontal, starting from rest. Calculate the speed the child will have at the foot of the chute, assuming no energy is lost during the descent.

3 If no mechanical energy is lost, a skate-boarder descending a straight 40 metre slope will arrive at the foot of the slope with speed $15 \, \text{m s}^{-1}$. Calculate the angle the slope makes with the horizontal.

4 A box of mass 20 kg is pulled up a ramp inclined at 30° to the horizontal. The work done in moving the box 10 metres is 1200 J. Calculate the magnitude of the average resisting force.

5 A cyclist free-wheels down a slope inclined at 15° to the horizontal, increasing his speed from $4 \, \text{m s}^{-1}$ to $10 \, \text{m s}^{-1}$ over a distance of 50 metres. Calculate the mean resistance on the cyclist, given that the mass of the cyclist and his bicycle is 60 kg.

6 A parachutist of mass 70 kg falls from a stationary helicopter at an altitude of 1 km. She has speed $8 \, \text{m s}^{-1}$ on reaching the ground. Calculate the work done against air resistance.

7 A tractor of mass 500 kg pulls a trailer of mass 200 kg up a rough slope inclined at 17° to the horizontal. The resistance to the motion is 4 N per kg. Calculate the work done by the tractor engine, given that the vehicle travels at a constant speed of $1.4 \, \text{m s}^{-1}$ for 2 minutes.

8 A box of mass 5.0 kg is pulled from A to B across a smooth horizontal floor by a horizontal force of magnitude 10 N. At point A, the box has speed $1.5 \, \text{m s}^{-1}$ and at point B the box has speed $2.8 \, \text{m s}^{-1}$. Find:

   a   the increase in kinetic energy of the box

   b   the work done by the force

   c   the distance AB.

PS 9 A car driver brakes on a horizontal road and slows down from $20 \, \text{m s}^{-1}$ to $12 \, \text{m s}^{-1}$. The mass of the car and its occupants is 1150 kg. Find the loss in kinetic energy. Given that the work done against resistance to motion is 50 kJ, find the work done by the brakes.

10 A bullet of mass 10 grams passes horizontally through a target of thickness 5.0 cm. The speed of the bullet is reduced from $240\,\text{m s}^{-1}$ to $90\,\text{m s}^{-1}$. Calculate the magnitude of the average resistive force exerted on the bullet.

9.2 Conservation of energy in a system of conservative forces

WORKED EXAMPLE 9.2

A child slides down a fun-fair ride. The child starts with speed $1\,\text{m s}^{-1}$ and descends through a vertical height of 4 metres. Find the speed of the child at the end of the ride if there is no resistance to the motion.

> **TIP**
>
> For a closed system of conservative forces the total mechanical energy, KE + GPE is constant.

Answer

Increase in KE	Mass of child $= m\,\text{kg}$.
$= \dfrac{1}{2} \times m \times (v^2 - 1^2)$	Final speed $= v\,\text{m s}^{-1}$.
Increase in GPE	Use PE $= mgh$.
$= m \times 10 \times -4$	As the child has descended, the
$= -40m$	increase in potential energy is negative.
Mechanical energy is conserved	As there are no resistances, increase in KE = loss in PE.
$\dfrac{1}{2}m(v^2 - 1) - 40m = 0$	Set up an equation to solve.
$v^2 - 1 = 80$	The mass of the child cancels in the
$v = 9$	equation.
Final speed $= 9\,\text{m s}^{-1}$	Solve to find the final velocity.

EXERCISE 9B

1 A particle is projected with speed $4\,\text{m s}^{-1}$ up a line of greatest slope of a smooth ramp inclined at $30°$ to the horizontal. It reaches the top of the ramp with speed $1.2\,\text{m s}^{-1}$. Calculate the length of the ramp.

2 A simple pendulum is modelled as a thread of length 0.7 metres, fixed at one end and with a particle (called the 'bob') attached to the other end. As the pendulum swings, the greatest speed of the bob is $0.6\,\text{m s}^{-1}$. Calculate the angle through which the pendulum swings.

3 A particle of mass 0.2 kg is attached to one end of a light rod of length 0.6 metres. The other end of the rod is freely pivoted at a fixed point O. The particle is released from rest with the rod making an angle of $60°$ with the upward vertical through O. Calculate the speed of the particle when the rod is:

   a horizontal   b vertical.

4 Two particles of mass 0.3 kg and 0.5 kg are connected by a light inextensible string passing over a smooth rail. The particles are released from rest with the string taut and vertical except where it is in contact with the rail. Calculate the velocity of the particles after they have moved 1.3 metres.

(PS) 5 Particles of mass 1.2 kg and 1.4 kg hang at the same level, connected by a long light inextensible string passing over a small smooth peg. They are released from rest with the string taut. Calculate the separation of the particles when they are moving with speed $0.5 \,\mathrm{m\,s^{-1}}$.

(PS) 6 A particle of mass 1.2 kg is at rest 2 metres from the edge of a smooth horizontal table. It is connected by a light inextensible string, passing over a light pulley on smooth bearings at the edge of the table, to a particle of mass 0.7 kg which hangs freely. The system is released from rest. Calculate:

a the distance moved by the particles when their speed is $3 \,\mathrm{m\,s^{-1}}$

b the speed of the particles just before the heavier particle reaches the pulley.

9.3 Conservation of energy in a system with non-conservative forces

WORKED EXAMPLE 9.3

A train of mass 80 000 kg is travelling at $15 \,\mathrm{m\,s^{-1}}$ when the brakes are applied. The train comes to a rest after travelling 2 km horizontally.

    a   Work out the loss in mechanical energy.

    b   Into what forms has this energy been changed?

    c   Calculate the average resistance force.

Answer

a Decrease in KE   Use decrease in $\mathrm{KE} = \frac{1}{2}mu^2 - \frac{1}{2}mv^2$.

$\quad = \dfrac{1}{2} \times 80\,000 \times 15^2$

$\quad = 9\,000 \,\mathrm{kJ}$

b Heat, sound and possibly some light.

c $9\,000\,000 = 2000\,F$ F = average resistance force. Use work done against resistance force = force × distance.

$\qquad\qquad F = 4500 \,\mathrm{N}$ Solve to find the average resistance force.

1   In an amusement park, a boy reaches the foot of a slide with speed $7\,\mathrm{m\,s^{-1}}$, after starting from rest. Because of friction, only 25% of his initial potential energy has been converted into kinetic energy. Calculate the vertical distance the boy has descended.

2   An object of mass $1.6\,\mathrm{kg}$ rests on a smooth slope inclined at $10°$ to the horizontal. It is connected by a light inextensible string passing over a smooth rail at the top of the slope to an object of mass $0.8\,\mathrm{kg}$ which hangs freely. After their release from rest, calculate:

   a   their speed when they have moved 0.5 metres

   b   the distance they have moved when their speed is $3\,\mathrm{m\,s^{-1}}$.

3   Two particles of mass $0.1\,\mathrm{kg}$ and $0.2\,\mathrm{kg}$ are attached to the ends of a light inextensible string which passes over a smooth peg. Given that the particles move vertically after being released from rest, calculate their common speed after each has travelled $0.6\,\mathrm{m}$. Deduce the work done on the lighter particle by the string, and use this to calculate the tension in the string.

PS 4   A particle of mass $2.2\,\mathrm{kg}$ rests on a smooth slope inclined at $30°$ to the horizontal. It is connected by a light inextensible string passing over a smooth rail at the top of the slope to a particle of mass $2.7\,\mathrm{kg}$ which hangs freely. The particles are set in motion by projecting the lighter particle down a line of greatest slope with speed $4\,\mathrm{m\,s^{-1}}$. Find the distance the particles travel before their direction of motion is reversed. Find the total energy gained by the hanging mass during this part of the motion, and hence find the tension in the string.

PS 5 A and B are two points $1\,\mathrm{km}$ apart on a straight road, and B is 60 metres higher than A. A car of mass $1200\,\mathrm{kg}$ passes A travelling at $25\,\mathrm{m\,s^{-1}}$. Between A and B the engine produces a constant driving force of 1600 newtons, and there is a constant resistance to motion of 1150 newtons. Calculate the speed of the car as it passes B.

6   A box of mass $80\,\mathrm{kg}$ slides down a curved path AB of length $20\,\mathrm{m}$. Point A is $3\,\mathrm{m}$ above point B. A constant resistance of $100\,\mathrm{N}$ acts throughout the motion. The particle starts from rest at A.

   Calculate the speed of the box when it reaches B.

9.4 Power

A car of mass $1500\,\mathrm{kg}$ accelerates up a hill against a resistance of $400\,\mathrm{N}$. The hill is modelled as a slope that is inclined at an angle θ to the horizontal, where $\sin\theta = 0.16$.

At a certain instant the engine is working at $60\,\mathrm{kW}$ and the car is travelling at $20\,\mathrm{m\,s^{-1}}$. Calculate the acceleration of the car at this instant.

TIP

Power = rate of doing work = Fv where F, the driving force, is constant.

Answer

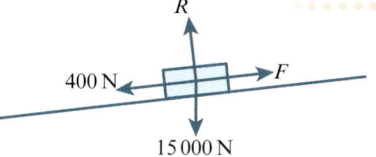

<table>
<tr><td>Draw a force diagram.</td></tr>
</table>

$60\,000 = 20\,F$ .... Power = driving force × speed.

$F = 3000\,\text{N}$ .... Solve to find the driving force.

Resultant force up slope
$\quad = F - 15\,000 \sin\theta - 400$:
$\quad = 3000 - 2800$
$\quad = 200\,\text{N}$

Resolve in the direction of motion to find the resultant force up the slope.

Newton's second law: .... Resultant force $= ma$.
$\quad 200 = 1500\,a$
$\quad a = 0.133\,\text{m}\,\text{s}^{-2}$ .... Solve to find the acceleration of the car at this instant in time.

EXERCISE 9D

 1 A helicopter of mass 800 kg rises to a height of 170 metres in 20 seconds, before setting off in horizontal flight. Calculate the potential energy gain of the helicopter, and hence estimate the mean power of its engine. State a form of kinetic energy that has been ignored in this model.

2 A 160 kg barrel of bricks is raised vertically by a 2 kW engine. Calculate the distance the barrel will move in 7 seconds travelling at a constant speed.

3 A car of mass 1800 kg, including the driver and passengers, is travelling at 32 m s^{-1} on a level road when the driver makes an emergency stop. The car comes to rest 5 seconds after the brakes were applied.

Find the power that the brakes provide.

4 A car engine has a maximum driving force of 750 N when travelling at 80 km h^{-1}. Calculate the power of the engine.

5 A train engine has a power rating of 2.5 MW. Calculate the tractive force when the train is travelling at 216 km h^{-1}.

 6 A boat is travelling at a constant speed of 16 km h^{-1}. The boat has mass 10.5 tonnes and the engine is working at its maximum power output of 8 kW. Calculate the work done when the boat is displaced $(5.0\mathbf{i} + 6.0\mathbf{j})$ km.

7   Alina is cycling on level ground. Alina and her cycle have a combined mass of 61.5 kg and she is working at a rate of 380 W. Given that Alina is accelerating at $0.65\,\mathrm{m\,s^{-2}}$, find the sum of the resistive forces acting on Alina and her cycle at the instant when her speed is $7.5\,\mathrm{m\,s^{-1}}$.

8   Victor is driving his 35 tonne truck on a horizontal road. Victor accelerates from $50\,\mathrm{km\,h^{-1}}$ to $65\,\mathrm{km\,h^{-1}}$, which is his maximum speed at 500 kW power output. Find the maximum acceleration of the truck, assuming that total resistance is constant.

9   Vince is driving his van against a constant resistance to motion of 5500 N. The van has mass 2.8 tonnes and engine power 125 kW. Vince's acceleration at the instant when his speed is $u\,\mathrm{m\,s^{-1}}$ is $0.90\,\mathrm{m\,s^{-2}}$. Calculate u.

P   10   The resistance to motion of a car is $kv^{\frac{3}{2}}$, where $v\,\mathrm{m\,s^{-1}}$ is the speed of the car and k is a constant. The power of the car's engine is 12 kW and the car has a constant speed of $27.5\,\mathrm{m\,s^{-1}}$ along a horizontal road. Show that $k = 3.0$.

11   A rocket, *Athena*, of mass 500 kg is moving in a straight line in space, without any resistance to motion. *Athena's* rocket motor is working at a constant rate of 500 kW and its mass is assumed to be constant. *Athena's* speed increases from $90\,\mathrm{m\,s^{-1}}$ to $140\,\mathrm{m\,s^{-1}}$ in time t seconds.

a   Calculate the value of t.

b   Calculate *Athena's* acceleration when its speed is $120\,\mathrm{m\,s^{-1}}$.

END-OF-CHAPTER REVIEW EXERCISE 9

1   A car travels along a straight horizontal road. The resistance to motion is kv N, where v is the speed of the car, measured in $\mathrm{m\,s^{-1}}$. The car travels at a constant speed of $25\,\mathrm{m\,s^{-1}}$ and the engine of the car works at a constant rate of 21 kW. Show that $k = 33.6$.

2   A van travels along a horizontal road. The resistance force is $20v$ N, where v is the speed of the van in $\mathrm{m\,s^{-1}}$.

The maximum speed that the van can achieve is $25\,\mathrm{m\,s^{-1}}$.

a   Show that when the van is travelling at a constant speed of $25\,\mathrm{m\,s^{-1}}$ the power being produced by the engine is 12.5 kW.

The mass of the van is 2000 kg.

b   Find the acceleration of the van when it is travelling at 60% of its maximum speed and the engine is working at 7.5 kW.

3   A sports car is travelling on a horizontal track. When the engine is working at full power, the car is travelling at a maximum speed of $80\,\mathrm{m\,s^{-1}}$ and the total resistance force is 2500 N.

Find the maximum power output of the engine.

4   A train, of mass 14 tonnes, moves along a horizontal track. A constant resistance force of 3600 N acts on the train. The power output of the engine of the train is 150 kW. Find the acceleration of the train when its speed is $30\,\mathrm{m\,s^{-1}}$.

5 Ahmed is driving a train of mass 65 tonnes at its maximum speed of $42.5\,\text{m}\,\text{s}^{-1}$ on a straight horizontal track. Its engine is working at $850\,\text{kW}$.

   a Find the magnitude of the resistance acting on the train.

When the train is travelling at $42.5\,\text{m}\,\text{s}^{-1}$, Ahmed disengages the engine and the train slows down. Assume that the total resistance is unchanged.

   b Use a work–energy calculation to find how far the train travels as it reduces its speed from $42.5\,\text{m}\,\text{s}^{-1}$ to $32.5\,\text{m}\,\text{s}^{-1}$.

6 A parcel of mass 2 kg slides down a rough inclined plane. The distance that the parcel travels is 5 metres and the parcel descends through a height of 1 metre. The parcel starts from rest and after 5 metres is at speed $3\,\text{m}\,\text{s}^{-1}$.

   a Calculate the average resistance force acting on the parcel.

   b Assuming that the resistance is constant, calculate the distance that the parcel has travelled when it reaches a speed of $1.5\,\text{m}\,\text{s}^{-1}$.

PS 7 A particle P of mass 0.2 kg is projected vertically upwards from ground level with speed $3\,\text{m}\,\text{s}^{-1}$.

At the same time a particle Q of mass 0.1 kg is dropped, from rest, from a point 12 m vertically above the launch point of P.

Each particle is subject to air resistance of 0.08 N.

The particles meet when P has risen through a height h m and is at its maximum height.

   a Use the work-energy principle to find h.

   b Find the speed of Q just before it meets P.

The particles then coalesce to form a single particle, R, of mass 0.3 kg.

   c Find the speed of R just after it has been formed.

Particle R is also subject to air resistance of 0.08 N.

   d Use the work–energy principle to find the speed of R when it reaches the ground.

8 A boy and his sledge have a total mass of 75 kg. The boy sits on his sledge at the top of a slope. He sets off down the hill with initial speed $2\,\text{m}\,\text{s}^{-1}$.

The boy travels 200 m and descends through 20 m. At the end of his journey the boy has speed $8\,\text{m}\,\text{s}^{-1}$.

   a Use the work–energy principle to find the average resistance acting on the boy.

   b Assuming that the resistance is constant, find the time taken for the descent.

9 Objects A and B, with masses $2.8\,$kg and $3.2\,$kg respectively, are attached to the ends of a light inextensible string which passes over a small, smooth pulley.

The system is released from rest with A and B at the same height above the ground.

a Find the tension in the string.

After 3 seconds of travel, A has not reached the pulley and B has just reached the ground.

b Find, for the motion in the first 3 seconds:

 i the gain in potential energy for A

 ii the work done on A by the tension in the string

 iii the gain in kinetic energy.

When B reaches the ground some of the kinetic energy is absorbed. B rebounds with 90% of the kinetic energy that it had when it hit the ground.

At the instant when B comes to rest, at the top of its path, A gets to the pulley.

c Find:

 i the height of the pulley above the ground

 ii the length of the string.

10 A truck of mass $3200\,$kg is driven along a horizontal road. The resistance to motion of the truck is constant and has magnitude $1000\,$N. The engine of the truck is working at a constant rate of $16\,$kW.

a Calculate the maximum possible speed of the truck.

b Calculate the acceleration of the truck at the instant when its speed is $10\,\mathrm{m\,s^{-1}}$.

The truck accelerates from rest until the maximum possible speed is achieved. It travels at this speed for a time and then the engine fails and provides no more driving force. The driver brings the truck safely to rest in a distance of $200\,$m.

c Find the work done by the brakes in bringing the truck to rest, assuming that the road remains horizontal.

11 A car manufacturer plans to bring out a new model with a top speed of $65\,\mathrm{m\,s^{-1}}$ to be capable of accelerating at $0.1\,\mathrm{m\,s^{-2}}$ when the speed is $60\,\mathrm{m\,s^{-1}}$. Wind-tunnel tests on a prototype suggest that the air resistance at a speed of $v\,\mathrm{m\,s^{-1}}$ will be $0.2v^2$ newtons. Find the power output (assumed constant) of which the engine must be capable at these high speeds, and the constraint which the manufacturer's requirement places on the total mass of the car and its occupants.

12 If the air resistance to the motion of an airliner at speed $v \, \text{m s}^{-1}$ is given by kv^2 newtons at ground level, then at 6000 metres the corresponding formula is $0.55 \, kv^2$, and at 12 000 metres it is $0.3 \, kv^2$. If an airliner can cruise at $220 \, \text{m s}^{-1}$ at 12 000 metres, at what speed will it travel at 6000 metres with the same power output from the engines?

Suppose that $k = 2.5$ and that the mass of the airliner is 250 tonnes. As the airliner takes off its speed is $80 \, \text{m s}^{-1}$ and it immediately starts to climb with the engines developing three times the cruising power. At what angle to the horizontal does it climb?

 13 A smooth wire is bent into the shape of the graph of $y = \dfrac{1}{6}x^3 - \dfrac{3}{2}x^2 + 4x$ for $0 < x < 6$, the units being metres. Points A, B and C on the wire have coordinates $(0, 0)$, $(3, 3)$ and $(6, 6)$. A bead of mass m kg is projected along the wire from A with speed $u \, \text{m s}^{-1}$ so that it has enough energy to reach B but not C. Prove that u is between 8.16 and 10.95, to 2 decimal places, and that the speed at B is at least $2.58 \, \text{m s}^{-1}$.

If $u = 10$, the bead comes to rest at a point D between B and C. Find:

a the greatest speed of the bead between B and D

b the coordinates of D, to 1 decimal place.

What happens after the bead reaches D?

 14 A block of mass M is placed on a rough horizontal table. A string attached to the block runs horizontally to the edge of the table, passes round a smooth peg, and supports a sphere of mass m attached to its other end. The motion of the block on the table is resisted by a frictional force of magnitude F, where $F < mg$. The system is initially at rest.

a Show that, when the block and the sphere have each moved a distance h, their common speed v is given by $v^2 = \dfrac{2(mg - F)h}{M + m}$.

b Show that the total energy lost by the sphere as it falls through the distance h is $\dfrac{m(Mg + F)h}{M + m}$. Hence find an expression for the tension in the string.

c Write down an expression for the energy gained by the block as it moves through the distance h. Use your answer to check the expression for the tension which you found in part **b**.

95

Answers

Answers to proof-style questions are not included.

1 Velocity and acceleration

Exercise 1A

1. 200 s
2. 72 km south
3. 25.2 km south
4. a i −410 m ii −290 m
   b i 290 m ii 210 m
   c i −120 m ii −210 m
5. a i 7.32 m s^{-1}, 2.93 m s^{-1}
   ii 8.64 m s^{-1}, −1.48 m s^{-1}
   b i 15.3 m s^{-1}, 0 m s^{-1}
   ii 9.55 m s^{-1}, 0 m s^{-1}
6. About $5\frac{1}{2}$ minutes
7. 8.18×10^{13} km

Exercise 1B

1. a i 1.00 m s^{-1}
      ii 17.2 m s^{-1}
   b i 18.7 km h^{-1}
      ii 0.936 km h^{-1}
   c i 0.00926 m s^{-2}
      ii 0.0347 m s^{-2}
   d i 10 600 km h^{-2}
      ii 35 000 km h^{-2}
2.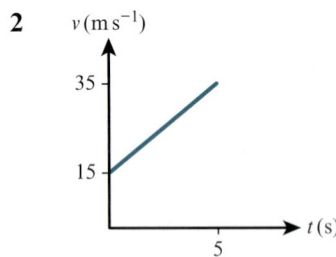

   4 m s^{-2}, 125 m
3. $7\frac{1}{2}$ m s^{-1}
4. 30 s, 2 m s^{-2}
5. 20 m s^{-1}, 1500 m
6. 25 s, 175 m
7. 40 s, $\frac{7}{8}$ m s^{-2}
8. 0.06 m s^{-2}

Exercise 1C

1. Proof
2. a Proof b 49.9 m s^{-1}
3. Proof
4. a 2.7 km b 250 s
5. 60 m
6. $(5 + 0.8t)$ m s^{-1}, $(5t + 0.4t^2)$ m; 15 s, 17 m s^{-1}
7. 4.43 s
8. 8.05 s
9. 66.7 m
10. a 5.29 m
    b It will go an unlimited distance in the negative direction.
11. $2\frac{1}{2}$ m s^{-2}, 2 m s^{-2}
12. $\left(\frac{1}{2}t^2 - 4t + 10\right)$ m; yes

Exercise 1D

1. a 50 metres
   b Away
   c Particle changes direction
   d The particle stops moving (it is stationary), 30 m from P
   e When $t = 50$
   f Decreasing
   g Speed is decreasing, velocity is increasing
   h 200 m
2. a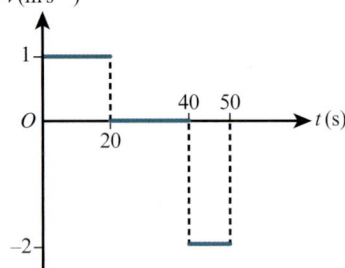

   b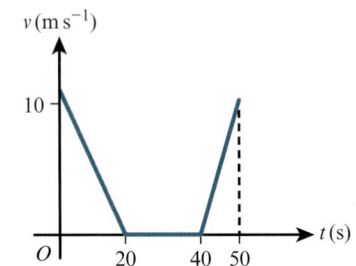

c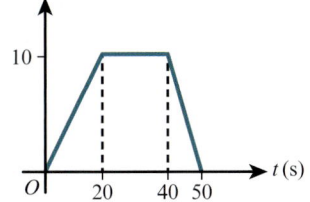

d $v\,(\mathrm{m\,s}^{-1})$

3 **a** The particle is stationary.

b 200 m, 6.67 m s^{-1}

c 13.3 m s^{-1}

4 $\dfrac{1}{32}\mathrm{m\,s}^{-2}$

5 **a** 50 m s^{-1}

b 5000 m

6 5D metres

7 s

8 **a** 71.8 metres

b Amir reaches the end of the 100 m before Sofia catches him.

9 **a** s

b 48 metres

10 **a** s

b 22 seconds

c 53.3 metres

11 −4

Exercise 1E

1 **i** **a** 0.65 m s^{-2}, −0.52 m s^{-2}

b 1950 m

ii **a** 1 m s^{-2}, −3 m s^{-2}

b 888 m

iii **a** −1.39 m s^{-2}, 1.25 m s^{-2}

b 398 m

2 **a** $d = 1350$ m

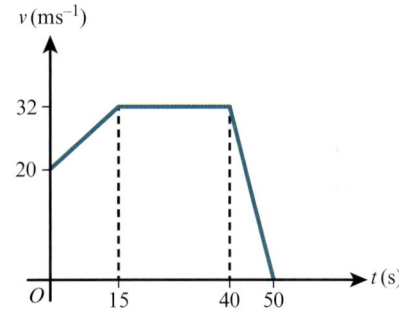

b $d = 495$ m

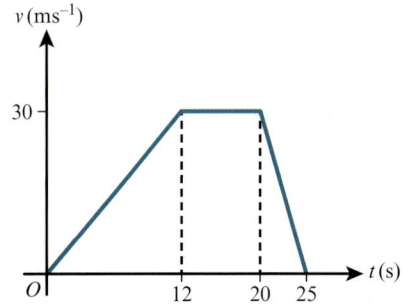

c $d = 863\,\text{m}$

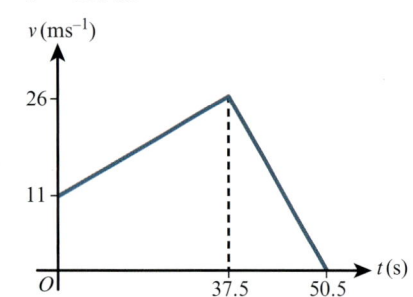

3 $1.6\,\text{m s}^{-2}, 60\,\text{m s}^{-1}, 15.5\,\text{s}$

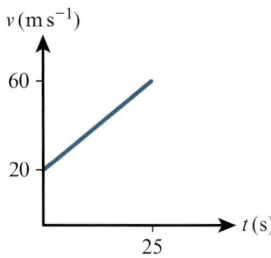

4 $6\,\text{s}, 1\frac{2}{3}\,\text{m s}^{-2}$

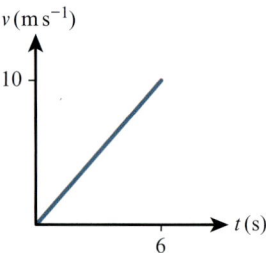

5 **a** $v\,(\text{m s}^{-1})$

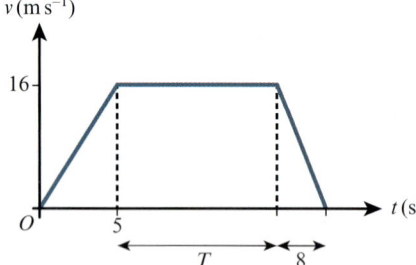

b 13
6 $22\,\text{s}$
7 $30\,\text{m s}^{-1}, 675\,\text{m}; 1\frac{1}{2}\,\text{m s}^{-2}; 20\,\text{s}$
8 $17.4\,\text{m s}^{-1}$
9 $12\,\text{m s}^{-1}, 44\,\text{s}$
10 $6.98\,\text{m}$

11 **a** $7\,\text{m s}^{-1}$ **b** $\dfrac{1}{2}$

     c $\dfrac{11}{28}$

12 44.2

Exercise 1F
1 $1.5\,\text{seconds}$
2 **a** $21.5\,\text{seconds}$ **b** $68.8\,\text{metres}$
3 **a** $8.1\,\text{metres}$ **b** $4\,\text{seconds}$
     c $4\,\text{bounces}$

End-of-chapter review exercise 1
1 **a** 0.4 **b** $2\,\text{km}$
2 **a** $2.5\,\text{m s}^{-2}$ **b** $10\,\text{s}$
3 $0.1\,\text{m s}^{-2}; 0.65\,\text{m s}^{-1}$
4 **a** $v\,(\text{m s}^{-1})$

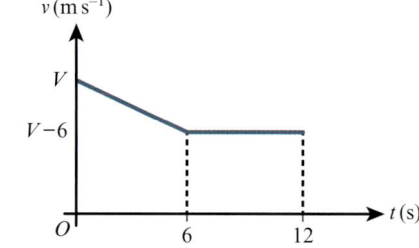

     b 19.5
5 $T = 6.8, V = 12.3$
6 **a** $360\,\text{s}$ **b** $4910\,\text{m}$
     c 92%
7 $0.2\,\text{m s}^{-2}; 0.338\,\text{m s}^{-2}$
8 $10.8\,\text{m}$
9 $0.08\,\text{m s}^{-2}$
10 **a** 25 **b** 40
11 Proof

2 Force and motion in one dimension

Exercise 2A
1 **a** $138\,\text{N}$ **b** $0.45\,\text{N}$
     c $7200\,\text{N}$ **d** $0.828\,\text{N}$
     e $0\,\text{N}$ **f** $1.18\,\text{N}$
     g $0\,\text{N}$
2 $1.5\,\text{m s}^{-2}$

3 $0.625\,\mathrm{m\,s^{-2}}$
4 $0.048\,\mathrm{N}$
5 $0.269\,\mathrm{s}$
6 $1.44\,\mathrm{m\,s^{-2}}$, $0.576\,\mathrm{N}$
7 $45\,\mathrm{m\,s^{-1}}$
8 $0.12\,\mathrm{m\,s^{-2}}$, $22.4\,\mathrm{s}$
9 $218\,\mathrm{N}$
10 $6.47\,\mathrm{kg}$

Exercise 2B
1 $1200\,\mathrm{kg}$
2 20
3 $2120\,\mathrm{N}$
4 115
5 $300\,\mathrm{N}$, $400\,\mathrm{N}$
6 $165\,\mathrm{N}$
7 52.2
8 $780\,\mathrm{N}$, $1\,\mathrm{min}\,20\,\mathrm{s}$
9 $2\,\mathrm{m\,s^{-2}}$
10 210

Exercise 2C
1 $17\,\mathrm{kg}$
2 $17\,900\,\mathrm{N}$
3 $13\,000\,\mathrm{N}$; $0.4\,\mathrm{m\,s^{-2}}$
4 $88.6\,\mathrm{kg}$
5 $1270\,\mathrm{N}$; $1.5\,\mathrm{m\,s^{-2}}$
6 $1280\,\mathrm{N}$
7 $s = \dfrac{Mv^2}{2(T - Mg)}$
8 $8400\,\mathrm{N}$

Exercise 2D
1 $2.8\,\mathrm{kg}$
2 $700\,\mathrm{N}$; the drum leaves the ground
3 $140\,\mathrm{kg}$
4 $67\,200\,\mathrm{N}$
5 The scales register the normal contact force between the man and the scales, not the 'weight'; the lift is decelerating at $0.444\,\mathrm{m\,s^{-2}}$.
6 80, 20
7 $53.6\,\mathrm{N}$
8 Between top and middle book $= 11\,m$
     Between middle and lowest book $= 22\,m$

End-of-chapter review exercise 2
1 **a** $863\,\mathrm{kg}$ **b** Lower mass
2 $11.9\,\mathrm{N}$
3 **a** $1.6\,\mathrm{s}$ **b** $1.23\,\mathrm{s}$
4 $4800\,\mathrm{N}$
5 $6530\,\mathrm{N}$
     a $40\,\mathrm{kg}$ **b** $7.5\,\mathrm{s}$
6 Proof
7 **a** There are forces acting upwards: buoyancy and water resistance.
     b 8.5, $0.255\,\mathrm{N}$
8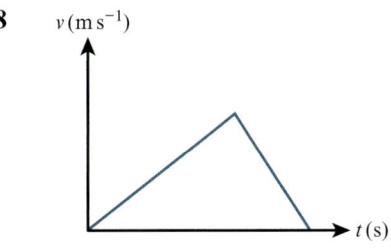
     $\dfrac{4}{3}mg$
     a $20\,\mathrm{m\,s^{-1}}$
     b $6\,\mathrm{s}$

3 Force in two dimensions

Exercise 3A
1 **a** $4\,\mathrm{N}$, $6.93\,\mathrm{N}$; $-7.88\,\mathrm{N}$, $-1.39\,\mathrm{N}$
     b $3.46\,\mathrm{N}$, $-2\,\mathrm{N}$; $-3\,\mathrm{N}$, $5.20\,\mathrm{N}$
     c $-4\,\mathrm{N}$, $0\,\mathrm{N}$; $2.05\,\mathrm{N}$, $-5.64\,\mathrm{N}$
     d $2.5\,\mathrm{N}$, $4.33\,\mathrm{N}$; $-2.5\,\mathrm{N}$, $4.33\,\mathrm{N}$
2 60, 8.66
3 **a** $41.8°$ **b** $112\,\mathrm{N}$
4 **a** Proof   **b** $68.2°$
     c 5.39
5 $5\,\mathrm{m\,s^{-2}}$, $10\,m\,\mathrm{N}$
6 Proof, $6\,\mathrm{kg}$
7 **a** $34.8°$ **b** $28.7\,\mathrm{N}$
8 $170\,\mathrm{N}$, $339\,\mathrm{N}$, proof
9 **a** $73.1°$ **b** 8.12
10 $\dfrac{240\sin\theta}{\sqrt{3}\sin\theta + \cos\theta}$, $\dfrac{120}{\sqrt{3}\sin\theta + \cos\theta}$

99

Exercise 3B

1   21.8 N, 63.9 N
2   35.2 N, 691 N
3   **a**   5.74 N
    b   7.00 N
    c   6.33 N
    In case **b** a component of the applied force will be added to $10\cos 35°$; in **c** it is subtracted.
4   559 N, 26.6°
5    $k = 252$ N, $a = 3.64\,\mathrm{m\,s^{-2}}$
6   **a**   They are two parts of the same string.
    b   4.53 N, 83.7°
7   52.4°
8   471, 4620

Exercise 3C

1   16.7°
2   19.3 N
3   0.580 kg
4   **a**   12.3 N, 11.3 N
    b   25.7 N, 37.6 N
5   4880 N, 8220 N
6   Proof; but then the angle between the first rope and the third rope is 180° and the component of T_2 perpendicular to the direction of T_1 has nothing to balance it.

Exercise 3D

1    $2.89\,\mathrm{m\,s^{-2}}$
2   166 N
3   8.05 kg, 85.7 N
4   Proof, $1.23\,\mathrm{m\,s^{-2}}$
5   986 N, 127 N
6   4.66 N, 28.0°, $1.73\,\mathrm{m\,s^{-2}}$
7   83.5, 74.9 N

Exercise 3E

1   099°
2    $1.59\,\mathrm{m\,s^{-2}}$
3   21.8°, 36.6°
4   354°
5   1.5°

End-of-chapter review exercise 3

1    $F = 4.62$, $G = 9.24$
2   34.6 N
3    $\theta° = 117°$, $F = 4.47$
4   **a**   73   **b**   0.009
5   **a** $6.5\,\mathrm{m\,s^{-2}}$ **b**   14.3°, 121°
6   **a**   Proof
    b $0.4\,\mathrm{m\,s^{-2}}$
    c   Slowed by friction and air resistance
7   0.7
8   **a**   7.11 N
    b   108 N

4 Friction

Exercise 4A

1   **a** $(P - Q)$ N in the direction of the force of magnitude Q N
    b $(Q - P)$ N in the direction of the force of magnitude P N
2    $80 \leqslant P \leqslant 120$
3   6.93 N
4   **a**   **i**   0.24   **ii**   0.005
    b   **i**   0.905   **ii**   0.1
5   9000 N
6   **a**   0.4   **b**   16.2 N

Exercise 4B

1   0.4, 63 N
2   0.25
3    $8.2 \leqslant P \leqslant 37.7$, to 1 decimal place
4    $2.62\,\mathrm{m\,s^{-2}}$
5   **a** $6.25\,\mathrm{m\,s^{-2}}$ **b**   0.625
6   **a**   Proof   **b**   48.3 N
7   Proof
8   Proof; $\alpha = \tan^{-1}\left(\dfrac{4}{3}\right) = 53.1°$

Exercise 4C

1   **a**   3.48 m
    b   0.870 s
    c   0.186 s
2   **a** $4.75\,\mathrm{m\,s^{-2}}$ **b**   3.66 m
3   0.024

4
  a $a = T - g$
  b $2.25g - T - F = 9a$
  c Up the slope with magnitude 5.73 N
  d Down the slope, with acceleration
     $0.652\,\mathrm{m\,s^{-2}}$

5
  a $10.5\,\mathrm{m\,s^{-2}}$
  b $14.1\,\mathrm{m\,s^{-2}}$
  c $5.71°$

6 $0.374; 1.58\,\mathrm{m\,s^{-2}}$

7 Proof

Exercise 4D

1
  a 0 N; the bowl remains at rest
  b 0 N; the bowl remains at rest with
     limiting equilibrium
  c 0.054 N; the bowl slides down with
     acceleration $0.108\,\mathrm{m\,s^{-2}}$

2 $24.2°$

3
  a 0.9   **b** $9.68\,m$

4 $\mu > 0.886$

End-of-chapter review exercise 4

1
  a $300 - X\cos\theta$ **b** $\dfrac{X\sin\theta}{300 - X\cos\theta}$

2
  a Proof   **b** 0.717

3
  a 0.32   **b** $11.3\,\mathrm{m\,s^{-1}}$

4
  a

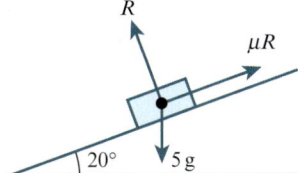

  b $0.13\,\mathrm{m\,s^{-2}}$
  c 0.35
  d No air resistance force acting; no other
     forces acting on the box; no turning
     effect (due to forces)

5
  a $\dfrac{1}{12}$ **b** $\dfrac{2}{3}$

6 Proof

7
  a $2\left(\mu\left(\sqrt{3} + 3\sqrt{2}\right) - 1 + 1.5\sqrt{2}\right)$ in
     direction of A moving up slope
  b Yes, 7.36 s after the start of the
     movement

5 Connected particles

Exercise 5A

1

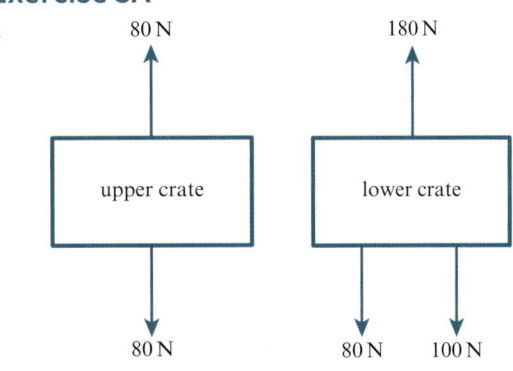

2
  a 200 N upwards
  b 200 N downwards
  c 1200 N upwards

3
  a

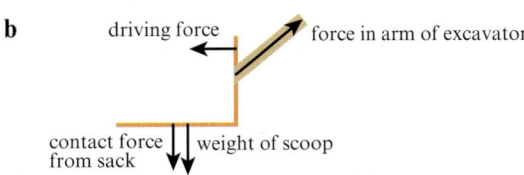

  b

4
  a

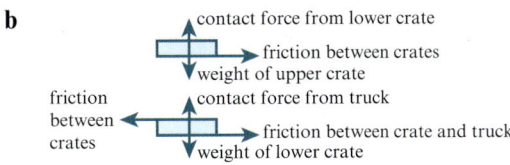

  b

  c

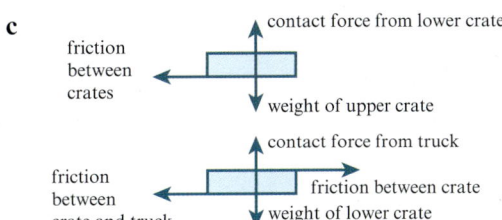

d   375 N

e   750 N

5   42

Exercise 5B

1   400 N, 500 N; 800 N forwards; it will appear that the car is being pushed from behind

2   **a** $12.7\,\mathrm{m\,s}^{-2}$

    b   4850 N

3   1120 N

4   **a**   10.8 kN

    b   5.5 kN, 2.75 kN

5   **a**   Driving (driving force = 2000 N)

    b   Tension of 1000 N between the engine and the first carriage, thrust of 1000 N between the two carriages

6   **a** $0.48\,\mathrm{m\,s}^{-2}$, 170 N

    b $0.08\,\mathrm{m\,s}^{-2}$, 170 N

7   250 N, 50 N

    a $0.05\,\mathrm{m\,s}^{-2}$ acceleration, 62.5 N

    b $0.05\,\mathrm{m\,s}^{-2}$ deceleration, 37.5 N

    c $0.156\,\mathrm{m\,s}^{-2}$ deceleration, 10.9 N

8 $\dfrac{F-(P+Q)}{M+m} + g\sin\alpha$; proof; proof

Exercise 5C

1   12.6 N, 22.1 N

2   37.5 N

3   **a**   312 N

    b   0.720 s

4   **a**   918 N, 540 N

    b   850 N, 500 N

5 $1.67\,\mathrm{m\,s}^{-2}$, 58.3 N

6   **a**   60 N

    b $2.86\,\mathrm{m\,s}^{-2}$, 42.9 N

7 $2.31\,\mathrm{m\,s}^{-2}$; 30.8 N, 7.69 N

8   3 : 1

9 $k = 5$

10   **a**   20 N

    b $1.25\,\mathrm{m\,s}^{-2}$

Exercise 5D

1   1248 N

2   1.20 N

3   **a**   5670 N

    b   371 N

4   **a**   1210 N

    b   850 N

5   345 kg

6   **a**   Up

    b   6700 N

End-of-chapter review exercise 5

1   **a**   0.849 s   **b**   0.424 s

2   **a**   0.5   **b**   20 m

3   **a** $\dfrac{20Mm\sin\theta}{M+m}$

    b $20\,\dfrac{(M-m)}{(M+m)}\sin\theta - 2\cos\theta$

    c   16.0

4   592 N

5   1.20 N

6   2340, 5850

7   **a**   2790 N

    b   1810 N

8   **a** $AB = 6.4\,\mathrm{N}$; $BC = 4.8\,\mathrm{N}$

    b   0.555

9   **a**   1140 N

    b   **i** $2.3\,\mathrm{m\,s}^{-2}$ downwards

        ii   4040 N

10   42.2 kg

11   **a**   23.2 N

    b   No (stops after total distance of 1.30 m)

12   **a**   1.41 s

    b   3.13 m

    c   2.12 s

6 General motion in a straight line

Exercise 6A

1   2.5 s and 3 s

2   **a** $19\,\mathrm{m\,s}^{-1}$ **b** $27\,\mathrm{m\,s}^{-1}$

3   5 s; 20 m

4   130 m

5   16 cm

6   **a**   30 cm

    b $24\,\mathrm{cm\,s}^{-1}$

    c $5.6\,\mathrm{cm\,s}^{-1}$

7   **a**   5 metres

    b   11 hours 25 minutes

    c $t = 8.57$

Exercise 6B

1 $4\,\mathrm{m\,s^{-1}}$
2 $v = 3t^2 + 4$, $a = 6t$; $22\,\mathrm{m}$, $16\,\mathrm{m\,s^{-1}}$, $12\,\mathrm{m\,s^{-2}}$
3 $1\,\mathrm{m\,s^{-1}}$, $-1\,\mathrm{m\,s^{-2}}$
4 **a** $0.1\,\mathrm{m\,s^{-1}}$, $1.2\,\mathrm{m\,s^{-2}}$
   b $1\,\mathrm{m\,s^{-1}}$
   c $1.92\,\mathrm{s}$, $3.00\,\mathrm{m}$
5 **a** $64\,\mathrm{m\,s^{-1}}$
   b
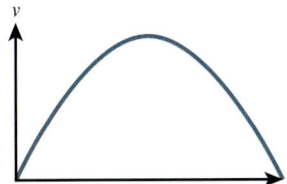
   c $48\,\mathrm{N}$
6 **a** $18\,\mathrm{m\,s^{-1}}$
   b $4\,\mathrm{m\,s^{-2}}$
7 **a** $120\,\mathrm{s}$, $1380\,\mathrm{m}$
   b $0.384\,\mathrm{m\,s^{-2}}$
   c $1.15\,\mathrm{m\,s^{-2}}$
   d $80\,\mathrm{s}$, $20.5\,\mathrm{m\,s^{-1}}$
8 **a** $952\,\mathrm{m}$, $62.4\,\mathrm{m\,s^{-1}}$
   b $-41.0\,\mathrm{m\,s^{-2}}$, $-10\,\mathrm{m\,s^{-2}}$, $-0.328\,\mathrm{m\,s^{-2}}$
   c $40\,\mathrm{m\,s^{-1}}$
   d $40\,\mathrm{s}$
   e
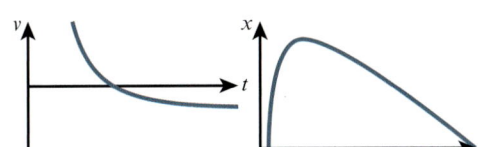

Exercise 6C

1 **a** $-3.6\,\mathrm{m\,s^{-2}}$
   b $0.96\,\mathrm{m\,s^{-1}}$
   c $x = 1.2t^2 - 0.5t^3$
2 $x = t^3 + 8t + 4$; $28\,\mathrm{m}$, $20\,\mathrm{m\,s^{-1}}$
3 $2\,\mathrm{m}$, $6\,\mathrm{m\,s^{-1}}$, $3\,\mathrm{m\,s^{-2}}$
4 $22\,\mathrm{m}$
5 $26\,\mathrm{m}$
6 $7.2\,\mathrm{m}$
7 **a** $54\,\mathrm{m}$
   b $4\,\mathrm{m}$
   c $4\,\mathrm{m}$

$62\,\mathrm{m}$; $54\,\mathrm{m}$
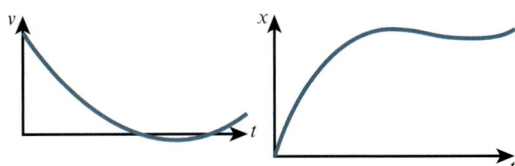

8 $28.9\,\mathrm{m\,s^{-1}}$
9 $1.82\,\mathrm{s}$
10 $10.2\,\mathrm{m\,s^{-1}}$

Exercise 6D

1 **a** $v = 0.04t^3 - 0.72t^2 + 4.32t$;
       $x = 0.01t^4 - 0.24t^3 + 2.16t^2$
   b $v = 8.64\,\mathrm{m\,s^{-1}}$, $x = 38.9\,\mathrm{m}$
2 **a** $v = 7.5 + t - 0.3t^2$
   b $x = 50\,\mathrm{m}$
   c $x = 62.5\,\mathrm{m}$
3 $-116\,\mathrm{m\,s^{-1}}$
4 $v = 10t - 2t^3 - 4$, $x = \dfrac{1}{2}(23 - 8t + 10t^2 - t^4)$;
   $11\frac{1}{2}\,\mathrm{m}$, $-4\,\mathrm{m\,s^{-1}}$, $10\,\mathrm{m\,s^{-2}}$
5 $12\,\mathrm{m}$, $7\frac{1}{3}\,\mathrm{m\,s^{-1}}$, $1\frac{7}{9}\,\mathrm{m\,s^{-2}}$
6 $60\,\mathrm{m}$
7 $28.5\,\mathrm{m}$
8 $4.99\,\mathrm{m\,s^{-1}}$, $5\,\mathrm{m\,s^{-1}}$
9 $1\frac{1}{2}\,\mathrm{m}$; Proof.
10 **a** $2\,\mathrm{s}$ **b** $8\,\mathrm{m}$
    c $6\,\mathrm{m\,s^{-2}}$
11 $74\,\mathrm{m\,s^{-1}}$, $282\,\mathrm{m}$
12 $\dfrac{3}{4}$, $205\,\mathrm{m}$

End-of-chapter review exercise 6

1 **a** when $t = 0$, $x = 0$, $\dfrac{\mathrm{d}x}{\mathrm{d}t} = 0$
   b $5.4\,\mathrm{m\,s^{-1}}$
2 **a** $-2.4\,\mathrm{m\,s^{-2}}$
   b **i** $13.5\,\mathrm{m}$ **ii** $40.5\,\mathrm{m}$
3 **a** 3.7 **b** $1.43\,\mathrm{m\,s^{-1}}$
   c $37.8\,\mathrm{m}$ **d** $7.13\,\mathrm{s}$
4 **a** Positive acceleration throughout,
       starting from rest
   b $3.13\,\mathrm{m\,s^{-1}}$

c

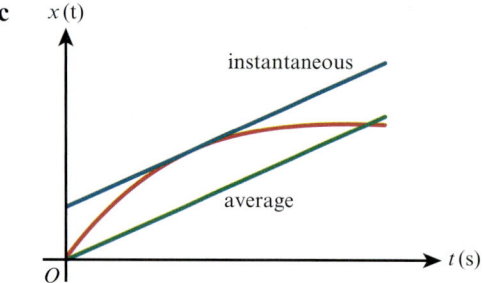

$x(t)$

instantaneous

average

O → $t\,(s)$

5 $0.24\,\mathrm{m\,s^{-1}}$, $0.4\,\mathrm{m}$

6   **a** $18\,\mathrm{m}$, $21\,\mathrm{m\,s^{-1}}$, $-24\,\mathrm{m\,s^{-2}}$
     b $3\,\mathrm{s}$, $-24\,\mathrm{m\,s^{-1}}$, $-6\,\mathrm{m\,s^{-2}}$
     c $1\,\mathrm{s}$, $28\,\mathrm{m}$, $-18\,\mathrm{m\,s^{-2}}$
     d $4\,\mathrm{s}$, $-26\,\mathrm{m}$, $-27\,\mathrm{m\,s^{-1}}$

7   **a** $0 < t < 2$
     b $1\frac{1}{4} < t < 2$ and $4 < t < 5$
     c $0.569\ldots < t < 2$ and $2.93\ldots < t < 5$

8   **a**   Proof
     b   3
     c   Proof
     d $1\frac{7}{9}\,\mathrm{m}$
     e $0 < t < 1$ and $t > \frac{3}{2}\sqrt{2}$

9 $5.86\,\mathrm{s}$, $34.1\,\mathrm{s}$

10   **a** $8.73\,\mathrm{s}$
     b   Proof
     c $0\,\mathrm{s}$

11 $34.7\,\mathrm{m}$

12   **a**   10
     b $B = 8A$, $C = A$
     c $27.1\,\mathrm{m\,s^{-1}}$

7 Momentum

Exercise 7A

1   **a** $2250\,\mathrm{N\,s}$ **b** $22.5\,\mathrm{N\,s}$
     c $0.2\,\mathrm{N\,s}$ **e** $28\,000\,\mathrm{N\,s}$
     d $20\,000\,\mathrm{N\,s}$

2 $900\,\mathrm{m\,s^{-1}}$

3 1.2 tonnes

4 4%

5 $24\,\mathrm{N\,s}$ away from the wall

6 $2500\,\mathrm{N\,s}$

Exercise 7B

1 $2.25\,\mathrm{m\,s^{-1}}$

2 $1.2\,\mathrm{m\,s^{-1}}$

3 $14\,\mathrm{m\,s^{-1}}$

4 $0.125\,\mathrm{m\,s^{-1}}$ in the direction that A was travelling before the collision

5 $0.96\,\mathrm{kg}$

6 $10\,\mathrm{m\,s^{-1}}$

7   **a** $v_1 = \dfrac{5m - 9}{m + 3}$, $v_2 = \dfrac{9 - 5m}{m + 3}$

     b $5\,\mathrm{kg}$ and $\dfrac{3}{7}\,\mathrm{kg}$

End-of-chapter review exercise 7

1 $2\,\mathrm{m\,s^{-1}}$ to $4\,\mathrm{m\,s^{-1}}$

2 $750\,000\,\mathrm{N\,s}$

3 $0.64\,\mathrm{m\,s^{-1}}$

4 19.5

5 $1.80 \times 10^6\,\mathrm{m\,s^{-1}}$

6 1.5

7 $2.5\,\mathrm{kg}$

8 $0.1\,\mathrm{m\,s^{-1}}$ away from BC (in the opposite direction to the original motion)

8 Work and energy

Exercise 8A

1 $6\,\mathrm{J}$

2 $125\,\mathrm{J}$

3   **a** $30\,000\,\mathrm{J}$ **b** $18\,000\,\mathrm{J}$

4 $700\,\mathrm{J}$, $0\,\mathrm{J}$, $700\,\mathrm{J}$

5 $4\,\mathrm{N}$

6 $23\,100\,\mathrm{J}$

7 $50\,\mathrm{N}$

8 $1130\,\mathrm{J}$

Exercise 8B

1   **a** $1620\,\mathrm{J}$ **b** $300\,\mathrm{kJ}$
     c $5.21 \times 10^6\,\mathrm{J}$ **d** $1600\,\mathrm{J}$
     e $6.4 \times 10^8\,\mathrm{J}$

2 $5.04\,\mathrm{kJ}$

3 $96.9\,\mathrm{kg}$

4 $37.8\,\mathrm{km\,h^{-1}}$

5 $12\,\mathrm{m\,s^{-1}}$

6 $240\,\mathrm{kJ}$

7 $400\,\mathrm{J}$

Exercise 8C

1. 200 kJ
2. 2.08×10^6 J
3. 3.5 m
4. **a** 20.25 J **b** 9 J
5. **a** 10 *mh* J **b** 10 *mh* J
6. **a** 0.776 m **b** 621 J

End-of-chapter review exercise 8

1. 32.7 N
2. **a** 1250 J **b** 1200 J
   c 47.1°
3. 60 J
4. **a** 5 m **b** 110 J
5. **a** 10 kJ **b** 144 kJ
6. **a** **i** 448 J **ii** 1344 J
   b 8 m s^{-1}
7. **a** The pebbles start with the same
      KE = 2.8 J
   b Raj's pebble gains speed and gains KE
      as it falls, assuming that the terminal
      velocity is not achieved.
   c Priya's pebble slows down as it rises and
      loses KE. It comes to rest with KE = 0
      and then falls downwards. When it is
      level with the starting point it has speed
      4 m s^{-1} downwards and KE = 2.8 J; from
      then on the KE is the same as it was for
      Raj's pebble.
8. **a** 800 J **b** 9 m s^{-1}
9. **a** 118 kJ **b** 96 kJ
   c 6 m s^{-1} **d** 2.7 metres
10. **a** 1000 J **b** 80 N

9 The work–energy principle and power

Exercise 9A

1. 40 J, 5 m
2. 12.2 m s^{-1}
3. 16.3°
4. 20 N
5. 105 N
6. 698 kJ

7. 814 kJ
8. **a** 14.0 J **b** 14.0 J
   c 1.40 m
9. 17 kJ; 97.2 kJ
10. 4950 N

Exercise 9B

1. 1.46 m
2. 26.0°
3. **a** 2.45 m s^{-1} **b** 4.24 m s^{-1}
4. 2.55 m s^{-1}
5. 0.025 m
6. **a** 1.22 m **b** 3.84 m s^{-1}

Exercise 9C

1. 9.8 m
2. **a** 1.48 m s^{-1} **b** 2.07 m
3. 2 m s^{-1}, 0.8 J, 1.33 N
4. 2.45 m, 44.6 J, 18.2 N
5. 13.2 m s^{-1}
6. 3.16 m s^{-1}

Exercise 9D

1. 1.36×10^6 J, 68 kW, the kinetic energy
   of the rotor
2. 8.75 m
3. 184 kW
4. 16.7 kW
5. 41.7 kN
6. 14 MJ
7. 10.7 N
8. 1.03 m s^{-2}
9. $u = 15.6$ m s^{-1}
10. Proof
11. **a** 5.75 s **b** 8.33 m s^{-2}

End-of-chapter review exercise 9

1. Proof
2. **a** Proof **b** 0.1 m s^{-2}
3. 200 kW
4. 0.1 m s^{-2}
5. **a** 20 000 N **b** 1220 m
6. **a** 2.2 N **b** 1.25 m
7. **a** 0.433 **b** 14.6 m s^{-1}
   c 8.42 m s^{-1} **d** 8.91 m s^{-1}
8. **a** 63.75 N **b** 40 s

9 **a** 29.9 N
   b **i** 84 J
      ii 89.6 J
      iii 5.6 J
   c **i** 6.15 m
      ii 6.3 m
10 **a** $16\,\mathrm{m\,s^{-1}}$
   b $0.188\,\mathrm{m\,s^{-2}}$
   c 210 J
11 54.9 kW, mass must not exceed 1950 kg
12 $180\,\mathrm{m\,s^{-1}}; 6.51°$

13 Proof
   a $6.83\,\mathrm{m\,s^{-1}}$
   b (5.7, 5)
      The bead retraces its motion back to A, with the same speed as before at each point of the wire.
14 **a** Proof
   b Proof, $\dfrac{m(Mg + F)}{M + m}$
   c $\dfrac{M(mg - F)h}{M + m}$